Leveraging Synergies Between Refining and Petrochemical Processes

Leveraging Synergies Between Refining and Petrochemical Processes

Eberhard Lucke and Edgar Amaro Ronces

CRC Press is an imprint of the
Taylor & Francis Group, an informa business

First edition published 2021
by CRC Press
6000 Broken Sound Parkway NW, Suite 300, Boca Raton, FL 33487-2742

and by CRC Press
2 Park Square, Milton Park, Abingdon, Oxon, OX14 4RN

© 2021 Taylor & Francis Group, LLC

CRC Press is an imprint of Taylor & Francis Group, LLC

Reasonable efforts have been made to publish reliable data and information, but the author and publisher cannot assume responsibility for the validity of all materials or the consequences of their use. The authors and publishers have attempted to trace the copyright holders of all material reproduced in this publication and apologize to copyright holders if permission to publish in this form has not been obtained. If any copyright material has not been acknowledged please write and let us know so we may rectify in any future reprint.

Except as permitted under U.S. Copyright Law, no part of this book may be reprinted, reproduced, transmitted, or utilized in any form by any electronic, mechanical, or other means, now known or hereafter invented, including photocopying, microfilming, and recording, or in any information storage or retrieval system, without written permission from the publishers.

For permission to photocopy or use material electronically from this work, access www.copyright.com or contact the Copyright Clearance Center, Inc. (CCC), 222 Rosewood Drive, Danvers, MA 01923, 978-750-8400. For works that are not available on CCC please contact mpkbookspermissions@tandf.co.uk

Trademark notice: Product or corporate names may be trademarks or registered trademarks and are used only for identification and explanation without intent to infringe.

Library of Congress Cataloging-in-Publication Data

Names: Amaro Ronces, Edgar, author. | Lucke, Eberhard, author.
Title: Leveraging synergies between refining and petrochemical processes / Edgar Amaro Ronces, Eberhard Lucke.
Description: First edition. | Boca Raton, FL : CRC Press, 2020. | Includes bibliographical references and index. | Summary: "This book provides a detailed description of the interfaces and connections between crude oil refining and petrochemicals. It offers a view of global and regional markets and economic opportunities for synergies between these sectors. The work serves as a reference that provides introductory and explanatory material as well as in-depth insight into future technology and market developments to help engineers and industry professionals understand the challenges and the potential benefits of developing expansion or optimization projects that may bridge the line between refining and petrochemicals"— Provided by publisher.
Identifiers: LCCN 2020043360 (print) | LCCN 2020043361 (ebook) | ISBN 9780367076771 (hbk) | ISBN 9780367649845 (pbk) | ISBN 9780429022043 (ebk)
Subjects: LCSH: Petroleum—Refining. | Petroleum chemicals.
Classification: LCC TP690 .A63 2020 (print) | LCC TP690 (ebook) | DDC 665.5/3—dc23
LC record available at https://lccn.loc.gov/2020043360
LC ebook record available at https://lccn.loc.gov/2020043361

ISBN: 978-0-367-07677-1 (hbk)
ISBN: 978-0-429-02204-3 (ebk)

Typeset in Times
by codeMantra

Contents

Preface

Writing a book takes a lot of dedication, patience, and hard work. You must be passionate about the story you want to tell; otherwise, you won't be able to finish it. It starts with the research and gathering of information, which to us was the fun part and could have lasted so much longer. But at some point, you must stop researching and gathering, and start analyzing, sorting, and aligning information along the train of thought you have for the book. At that time, you may have new ideas and change the concept you had laid out for each chapter to accommodate additional information or to compensate for lack of other information. Either way, it is easy to get lost in the process of putting the pieces together and of finalizing the chapters. Having the final goal in mind and knowing what you are trying to achieve helped us with staying focused on the task at hand, which was simply to write a book.

So, how do you know what potential readers might be interested in? To be honest, you don't. Yes, you ask around and do a survey to make sure there is some level of interest. But in the end, you believe that if you – the authors – find the topic interesting and want to know more about it, there will be others, with similar interests, with similar jobs, or with similar motivation to learn more about the topic. And these are the people you write the book for. But you try to write it in a way that it will be helpful for others as well. We live in a world that is subject to constant change. New technologies influence the way we communicate, the way we move from point A to point B, and the way we solve problems. Once you get used to something, it is outdated and replaced by something better. Technology has improved the speed in which we can do things, and the comfort from which we can do things. And sometimes, it is necessary to stop and take a high-level look at the mechanisms and technologies that allow us to do all these things that we have accepted as normal. We believe that understanding the source of our comfort and what it depends on allows for a much better appreciation of where we are today and how we got here.

The industry we both, Edgar and I, work in has a bad reputation and is blamed for a lot of bad things that are happening around us. And we are not saying that it is wrong to point these issues out and try to find a better solution. We just want to raise awareness of the fact that a lot of good things come out of the industry as well, and that it would be wrong and naïve to believe we could just walk away from it and do something else. It takes time and patience and the same innovative spirit that got us to where we are today to get us where we need to be in the future.

Writing a book and spreading information are only two of the many things that need to happen in order to focus the right minds with the right tools to work on the right issues. There is so much more that needs to be done. But it is a start, and we hope that we can do our part in the whole process to build a solid basis for the energy revolution that needs to take place and for everybody involved in the process to understand the opportunities we have at hand.

Eberhard Lucke
CEO – Lucke Consulting Technology Services, LLC

Edgar Amaro Ronces
Project Manager Americas Division – Vopak

Authors

Eberhard Lucke is a well-established and recognized expert in the oil refining and gas processing industries with more than 29 years of experience as a consultant, engineer, and manager. Starting as a process and technology engineer in one of Germany's leading refineries, he moved on to work in Venezuela, the USA, and South America in distinct roles such as consultant, lead process engineer, owner's engineer, project manager, department manager, and company executive. He led studies and design work in oil refining and petrochemicals, as well as gas monetization and carbon capture and utilization.

In 2012, Eberhard founded Lucke Consulting Technology Services, LLC, to utilize his technical expertise and industry experience in the growth and development of clients across the energy sector. Since then, he has supported many projects as owner's engineer and consultant by providing design reviews, due diligence, and technical management. He serves on the advisory board for the Professional Science Master's Program (Environmental) at Rice University and several other advisory boards.

Edgar Amaro Ronces is a chemical engineer who graduated from the National Autonomous University of Mexico and has more than 15 years' experience in the O&G industry. In his career, he worked for several EPC companies such as Fluor, Technip, and CH2M Hill and executed work tasks to accomplish the various stages of the design and construction process. In his roles, he has been involved in the design of several refining and petrochemical facilities. He is currently working as a project manager for the Americas Division of Vopak.

1 Introduction

Many people have expressed their understanding of the meaning of the word "synergy," and they all come up with a very good definition of synergy that reflects what the word means for them and their lives. One of my favorite quotes is the following:

> Synergy is better than my way or your way. It's our way.
>
> **Stephen R. Covey**

What I like about this quote is its simplicity and that it allows enough freedom to apply the concept to all situations where synergy might exist. Like many other words we use today, synergy comes from the Attic Greek. The Greek derived the term "synergia" from the word "synergos," which translates to "working together." Working together or teamwork is the basis for synergy, as it creates a new entity or team that is greater or of more value than the simple sum of its parts.

Consequently, there is no synergy in mathematics. $2+2$ will always be 4. There is no case to be made in which $2+2$ may result in a value greater than 4. Luckily for us, this logic doesn't apply to all disciplines, and especially when it comes to humans and human behavior, other rules apply. In a more contemporary definition, synergy is the interaction of two or more entities, for example, individuals, substances, or organizations, to produce a combined effect or outcome greater than the sum of their separate effects or outcomes. Based on this definition, synergy can be used as a synonym for collaboration, cooperation, or teamwork and vice versa.

A simple example for synergy would be coming from the world of team sports such as volleyball. Let's assume there was no teamwork in volleyball, and everything would be reduced to individual actions. One individual would serve the ball over the net, hoping for a mistake on the other side. The individual on the other side who is closest to the ball's landing point would play the ball back over the net, again hoping for a mistake on the other side. And this would continue until one individual will make a mistake by missing the ball, hitting it into the net or outside of the side lines. It is easy to see that volleyball without teamwork would be predictable, unattractive, and even boring. The fact that the team members work together and that the ball can be played up to three times on each side allows teams to utilize specific skills of different players in their specific positions to develop a variety of defending and attacking strategies to keep the opposing team guessing and to increase the chance of scoring points by a factor that is greater than the sum of the individual odds of scoring a point. These variations in approaching the game and the application of multiple strategies are what make the sport attractive for players as well as spectators. Synergies between players make the team much stronger and more successful than a group of individuals alone. Sometimes, this effect can be observed in games where a supposedly stronger team are the favorites in playing against a so-called underdog, a team of lesser quality. The assessment of quality in this case is based on the individual capabilities of each player. The favorite team would have star players with great individual skills that have reached fame and global recognition for their achievements. However, on any given day when these favorites don't play as a team and solely rely on the skills of their star individuals, an underdog can utilize the synergies of teamwork and beat the favorites.

This concept can easily be applied to work teams and businesses as well. We all know that in a team of colleagues and partners who all bring different skills and knowledge to the table, we can be much more effective and successful than as individuals in delivering good solutions and high-quality work on a consistent basis to our clients. Of course, there are always exceptions to the rule, and we all have had situations where it was easier to just get something done alone than to involve a whole team in the process. But these really are exceptions as they apply to only small tasks that can be done by an individual with the right skills and experience. Most of the time, we are facing challenges and tasks that are greater than that and that require much more than the effort of one person to get it done.

Going back to the analogy of team sports, there are certain conditions that must be satisfied to make synergy happen:

- Self-respect: each player must have respect for themselves and be confident in their own capabilities and value that they bring to the team.
- Respect for others: each player must respect the other players and recognize their value and contribution to the team; it also includes respect for the opponents and their qualities as they need to be considered in developing a match strategy.
- Communication: since we can't read each other's minds, efficient communication is key in bringing the team together and aligning for the common goal. Communication must be open and honest.
- Passion: each player must have the will and desire to win the game.

With these conditions in mind, we will look at two industries that have coexisted in the past and that had interactions to a certain degree, in some cases to a considerable level of integration: petroleum refining and petrochemicals. Petroleum refining is the main source of fuels and other products that drive our economies. Petrochemicals are the basic building blocks for most products that make our lives so much more comfortable and luxurious. For various reasons that will be discussed in this book, refining and petrochemicals are on the path toward a much higher degree of integration during the next 15 years. The integration will happen in existing facilities as well as be realized in new, integrated grassroots petrochemical and refining complexes. And in order to make these projects successful, all companies involved must ensure that the conditions for taking advantage of the synergies of such integration are fulfilled.

1.1 SELF-RESPECT

All companies or entities involved in the integration of refining and petrochemicals must understand their value and position within their respective markets, and their capabilities and strengths that they bring to the integrated team.

1.2 RESPECT FOR OTHERS

All companies or entities involved must have respect for each other and recognize each other's value in the integrated organization. This includes trust and confidence in the capabilities of each team member and their intention to make the integration successful. It also means that the integrated team has respect for their competitors in the respective markets and knows of their strengths and values as well.

1.3 COMMUNICATION

It must be said that in any relationship, open and honest communication is key to identifying and removing roadblocks, to alignment of teams, and to continuously assess and improve performance of all team members and the whole team. Only by applying trust and honesty can the team develop to the next level of performance. Communication makes it all happen. All members must know what the goals are. All members must know the plan for achieving the goals. All members must know the status of the team in the process and the challenges that have to be overcome. If communication leaves any room for uncertainty and guesswork, the process is determined to fail. There is no place for "sugar-coating" issues as this will reduce the likelihood of them being resolved. And in that context, the correct form of communication must be selected in each instance as personal discussions, meetings, emails, or phone calls all have their place in communication, but don't have the same effect and impact on the message that must be transmitted.

1.4 PASSION

All parties involved in the integration, from the refining side, from the petrochemicals side, and from any other supporting organization, must be aligned in the desire and will to achieve the newly defined, common goals of the integrated organization. The path to success is not easy, and there will be several hurdles to be taken, roadblocks to be removed, and resistance to be overcome. All these activities require energy and motivation in each individual, and the best source of energy and motivation is the passion for the cause people are working for. If that passion and commitment are not given in the whole organization, it will be more difficult, in some cases even impossible, to realize all potential synergies and reach the level of competitiveness that will be required to survive in a difficult market.

Each member of the integrated team must be open to change, open to new ideas, and flexible enough to adjust to new strategies and plans that will be developed. And these strategies and plans will not be set in stone forever. One of the keys to survival will be the ability to quickly adjust to changes in the market, no matter if the changes are supply- or demand-driven. In some cases, consumer habits will change and force adjustments to the manufacturing strategy and market position. This requires the right mindset of openness and flexibility and the strength to let go of what feels familiar and safe. The members of the integrated team must be willing to step out of their comfort zone and learn new ways of doing business.

Leveraging synergies and growing value does not always go with "going bigger" as might be thought based on the definition we used in the beginning of this introduction. For example, the increased value for the integrated refining and petrochemical complex in the area of common use of infrastructure and administration includes the effective and successful execution of certain functions such as financing, procurement, supply chain management, or product handling which can be achieved by an integrated team that is smaller than the sum of the two individual teams before integration. The value comes from the reduction in expenses, which directly reduces fixed expenses and increases the bottom line of the operation. And this makes the integrated team more competitive on the market. In an ideal world, there would be opportunities for the free resources coming from these teams to be integrated into other functions, but in many cases, this will be difficult as not all skills and knowledge can be transferred to other departments and disciplines. Integration is not all about happy and motivated employees; it is also about making tough decisions that are best for the organization.

The refining business is facing pressure from stagnating and even declining fuel markets in parallel to being hit by more restrictive environmental regulations and the decline of prestige of fossil fuels in the public.

The petrochemical industry sees an increase in competition on almost all markets as new players arise. Companies are looking for ways to become more efficient and competitive while also trying to stabilize their supply chains and compensate for the cyclic effects of supply and demand swings.

The integration of refining and petrochemicals is already and will be the preferred strategy of the players in the energy and chemical markets as it allows companies to take advantage of the synergies that can be unlocked by operating both businesses as one.

2 Fundamentals

2.1 HISTORICAL DEVELOPMENT OF PETROCHEMICAL-REFINING SYNERGIES

For many years, oil has provided feedstocks for fuels and petrochemicals; however, the needs driving each industry have been historically out of synchrony between themselves. First developments in petroleum refining were promoted by the need of lighting and heating without the intention of producing chemical feedstocks, which in those years were provided from some non-hydrocarbon process such as the by-products generated from steel production, fermentation, and some other natural resources rather than oil.

However, once the refining industry was pushed by an increased demand for fuels to provide the automobile industry with more efficient fuels, the cracking of heavier oil cuts to produce gasoline opened a new horizon for the production of chemicals since the naphthas and other light by-products from cracking and fractionation were rich in olefins and aromatic components, making these attractive as feedstocks for the production of chemicals and plastics. This is possibly the first sign of a synergy between oil refining and petrochemical industries.

Early development of chemicals production was promoted by the need to supply synthetic fibers, plastics, and fertilizers. One of the most relevant technologies was the production of ammonia from its basic elements, nitrogen, and hydrogen through the well-known Haber process, which broke a huge barrier on technology research using high-pressure equipment and catalysis development for its successful implementation. But its importance was beyond the technical challenges. In terms of development, the production of ammonia through this route made the natural gas a key resource in the petrochemical industry.

Following the success of the Haber process, natural gas was found to be an attractive source to produce petrochemicals, through an innovative process to produce ethylene, one of the building blocks of the petrochemical industry. Initially, the ethylene was mainly pursued as a precursor of ethylene glycol needed as antifreeze. The technology for production of plastics was still in development, and most of them relied on different sources as feedstock; for example, PVC was produced from chlorine and acetylene obtained from coal.

During the early 1900s, more chemical processes were successfully implemented, for example, the process of Nylon® and other fibers and plastics in low scale. However, from the commercial perspective, the growth of this industry was defined by the political and economic environment. All this research was taking place post WWI, during the Great Depression in the 1930s. Hence, the demand of durable products decreased substantially, and technology development was driven by the possible needs of supplies in a future war.

In that sense, Europe played an important role on technology development, and countries such as Germany were looking for self-sufficiency. Therefore, a good amount of infrastructure to produce fuels and synthetic rubber for their armies was built. However, in the search for independency, these developments were based on coal instead of petroleum.

After WWII, all the plants would no longer be needed to provide war supplies; therefore, many of these products could be commercialized for consumers, and once these products became available responding to several basic needs, the demand increased at a considerable high pace, changing the chemical industry where oil and natural gas replaced most of the non-hydrocarbon sources.

On the other hand with the economic boom post war, the automobile industry enjoyed a similar growth as chemical products or even better, and the new drivers demanded faster and more efficient vehicles. This requirement was transferred to the refining industry which had to respond with the production of improved fuels to meet the performance of new motors. In order to provide such products, new technologies such as catalytic cracking, hydrotreating, and aromatics extraction were developed.

Catalytic cracking started with the development of the fixed-bed process which produces high octane gasoline from heavy oil cuts with light olefins as by-products, using a reactor operated in semi-continuous mode, since the reactor had to shut down for regeneration of the catalyst. This process was superseded by an improved design able to work in synchrony with a regenerator in a fluidized catalytic bed.

Another key development was the reforming of naphthas, which produced a cut in the gasoline boiling range with branched and aromatic components. The product called "reformate" could be further treated in subsequent processes to recover the aromatic components, benzene, xylene, and toluene. This process represents some of the most important windows between the refining and petrochemical industries through the value of gasolines, olefins, aromatics, and hydrogen.

2.2 EVOLUTION OF KEY TECHNOLOGIES

2.2.1 Syngas Production

One of the cornerstones of the chemical and energy industry is the synthesis gas (syngas), and its importance goes beyond the value of the immediate feedstocks generated from its processing; it can be considered as a bridge between resources such as natural gas, coal, or crude oil with chemicals, fuels, and even power generation. Virtually, this bridge offers a two-way path among different sources/destinations, but in practice, there are many challenges to overcome. Hence, the research on process based on synthesis gas has received much attention for several years.

The syngas is a mixture of hydrogen and carbon monoxide with several uses from chemicals to fuels or power, and can be produced from several processes such as reforming of hydrocarbons, oxidation of coal/coke, biomass gasification, etc. In particular, for the oil and gas industry, the steam reforming of natural gas is the most mature technology with extensive research behind, and this has been driven by the needs of ammonia and its derivatives demanded by the fertilizer industry but also by the production of hydrogen required in refineries, to remove sulfur in fuels to meet environmental restrictions.

The ammonia production process is one of the most diverse technologies in terms of the individual unit operations involved. Therefore, many parameters were susceptible to be improved from their initial developments of the technology to arrive at the processes as known in these days.

Several challenges were successfully overcome during this journey; for example, it was found that sulfur and coke had a negative effect on the catalyst. Hence, adequate pretreatment of the feed stream was incorporated. Catalyst development was focused not only in the actual kinetics of the process but also in inhibition of side reactions, for example, to reduce the coking in the reactors.

Ammonia is synthesized from its basic components (nitrogen and hydrogen) through the Haber process which according to Le Chatelier's principle requires high pressure to increase the conversion to ammonia as illustrated in Figure 2.1.

Hence, during early developments of the process at industrial scale, the syngas (produced at lower pressures) was compressed using reciprocating machines which represented a considerable part of the capital and operational cost of these facilities. This particular issue pushed the research efforts to find an efficient way to carry out the steam reforming at higher pressures. Since also high temperatures were involved, significant improvements on the selection of mechanical materials were incorporated.

Incorporation of most efficient machines such as centrifugal compressors was justified only with increased demand and therefore the need of larger ammonia units, but once those were incorporated, a substantial improvement in efficiencies was achieved not only from the mechanical point of view but also in terms of thermodynamics. Because with the improved reforming process, heat was available and recovered producing steam which in turn could be used to run the compression system.

Several feedstocks were explored for the production of syngas with particular preference of liquid streams to replace the use of compressors with pumps, but in the end, the use of natural gas was predominant to allow not only ammonia production but also hydrogen. A typical configuration of the syngas process for ammonia production is shown in Figure 2.2.

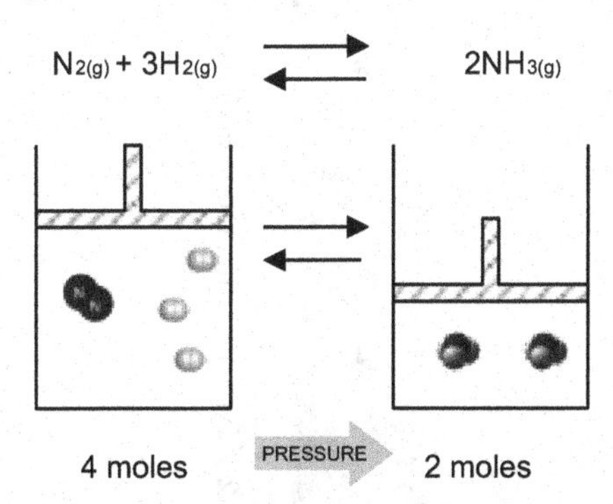

FIGURE 2.1 Effect of pressure to displace ammonia reaction toward less volume (moles) of gas.

FIGURE 2.2 Typical configuration of an ammonia production unit.

2.2.2 Ethylene Production

Ethylene is one of the main precursors of petrochemical products due to its high reactivity only preceded by acetylene with a similar chemical structure but more unstable (and therefore more reactive). In fact, initial efforts to produce chemicals were focused on acetylene rather than ethylene, taking advantage of early developments to produce acetylene from calcium carbide. However, the high reactivity of acetylene was not always a positive attribute, and the increased risk in their handling along with the high costs to source calcium carbide for its production led to the search for better alternatives.

Another option was found in the electric arc process which had the advantage of producing not only acetylene but also ethylene. Unfortunately, the electric cost became a great limitation on the development of this route; however, this research brought the ethylene into the picture and its chemical similarities with the acetylene drove the attention in the research, making the ethylene a potential candidate to generate the same downstream products in a cheaper and safer way.

Before light hydrocarbons and natural gas became more available, the first attempts to produce ethylene relied on coke ovens, and ethanol was also an ethylene precursor, but the increased demand of ethanol for human consumption along with further availability of ethane from natural gas reverted the commercial preference of this reaction being ethanol one of the derivatives of ethylene that drove the development of olefins industry.

Such availability of ethane in the natural gas around the USA had a great influence to promote the research on thermal cracking as an alternative to produce ethylene, but in the beginning, the main challenge was to find suitable purification processes at low temperatures to recover ethane from natural gas as well as to separate the resulting products from this reaction, mainly to recover ethylene from uncracked ethane. Fortunately, the unborn olefins industry found a solution on previous research of cryogenic liquefaction to produce oxygen from air, but one of the main targets has been to lower operational costs of refrigeration.

Based on a similar concept, the cracking of heavier liquid feedstocks followed the steps of ethane crackers, mainly in Europe and Asia where availability of light naphtha from refineries was more abundant than ethane from natural gas, but in this case in addition to ethylene, considerable amount of propylene and benzene became available, thus creating the foundation of the petrochemical industry.

The core of an olefin cracker is the pyrolysis reactor where the rupture of paraffinic hydrocarbon components takes place. Several parameters have to be optimized to achieve efficient levels of operation, but in the case of thermal cracking, temperature control and residence times and partial pressure are the most important as depicted in Figure 2.3.

Due to the high reactivity of ethylene, once it is produced, it can be easily converted into acetylene or even further degraded to coke, the ideal temperature to get efficient ethylene yields is in the order of 900°C. However, early developments of cracking furnaces had metallurgic limitations. Therefore, initial attempts explored the option of alternate cracking routes with increasing yields to acetylene. However,

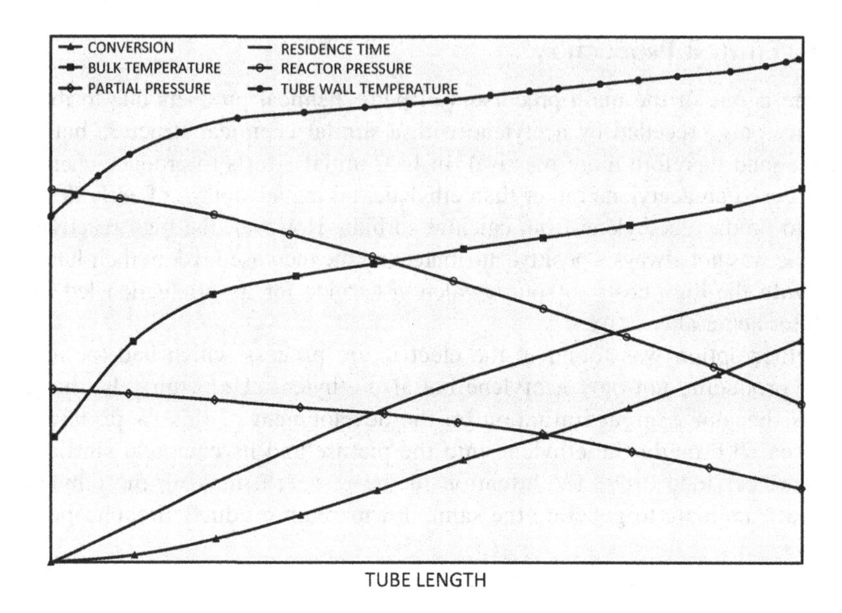

FIGURE 2.3 Kinetics of cracking for olefin production.

the production of olefins derivatives was more economically favorable from ethylene rather than acetylene. Catalytic routes that may require lower conversion temperatures were never materialized probably because of their limited amount of by-products that made the thermal cracking a more feasible process.

Hence, further developments of cracking furnaces were focused not only on better metallurgies to achieve higher temperatures but also on getting high conversion rates (cracking severity) with low residence times to inhibit the production of coke.

Pyrolysis reactions are exothermic, and therefore, a substantial amount of heat can be recovered. In typical reactor configurations, this heat recovery is achieved through steam generation. Furnace designs have evolved to a staged configuration, where in a first step, the feed stream is heated by the flue gases produced in the same combustion of the furnace, and this has been commonly referred to as the convection section which is followed by a radiation section where the heat comes from the direct flame in the furnace. One of the targets in the design of the reaction furnaces is to achieve an optimum layout to maximize the heat recovery between preheating and steam production. In the convection section, the energy from the flue gas is used to preheat the feed stream at above 500°C as well as the Boiler Feed Water (BFW) which will be further vaporized to produce steam, while in the radiation section, the feed stream reaches a temperature in the order of 800°C required for an adequate pyrolysis. Then, in order to control the extent of further undesired reactions, the temperature of product outlet has to drop suddenly, and hence, high-pressure steam is generated to release this massive amount of heat.

One of the greatest challenges superseded in the development of this technology was to find an integrated design of the whole reaction system including furnaces, and pre- and post-heat exchangers as one single unit to achieve an optimum design in terms of energy, but such design required a complex integrated mechanical structure, and its design represented also a challenge in terms of constructibility, a typical configuration of this system is shown in Figure 2.4.

As previously mentioned, the separation processes to recover high-purity products represented another challenge for this technology. Since components such as methane or ethane have very low boiling points, the use of cryogenic fractionation was needed. Different schemes have been explored aiming for an optimum configuration to reduce energy, and capital costs of a typical separation sequence are illustrated in Figure 2.5.

Regardless of the feedstock available, the cracking process implies the rupture of single C--H (carbon–hydrogen) bonds releasing a considerable amount of hydrogen that can be also recovered in this cryogenic stage. Yields of hydrogen go from 1% to 5% depending on the feed stream as shown in Figure 2.6. For example, for an ethane

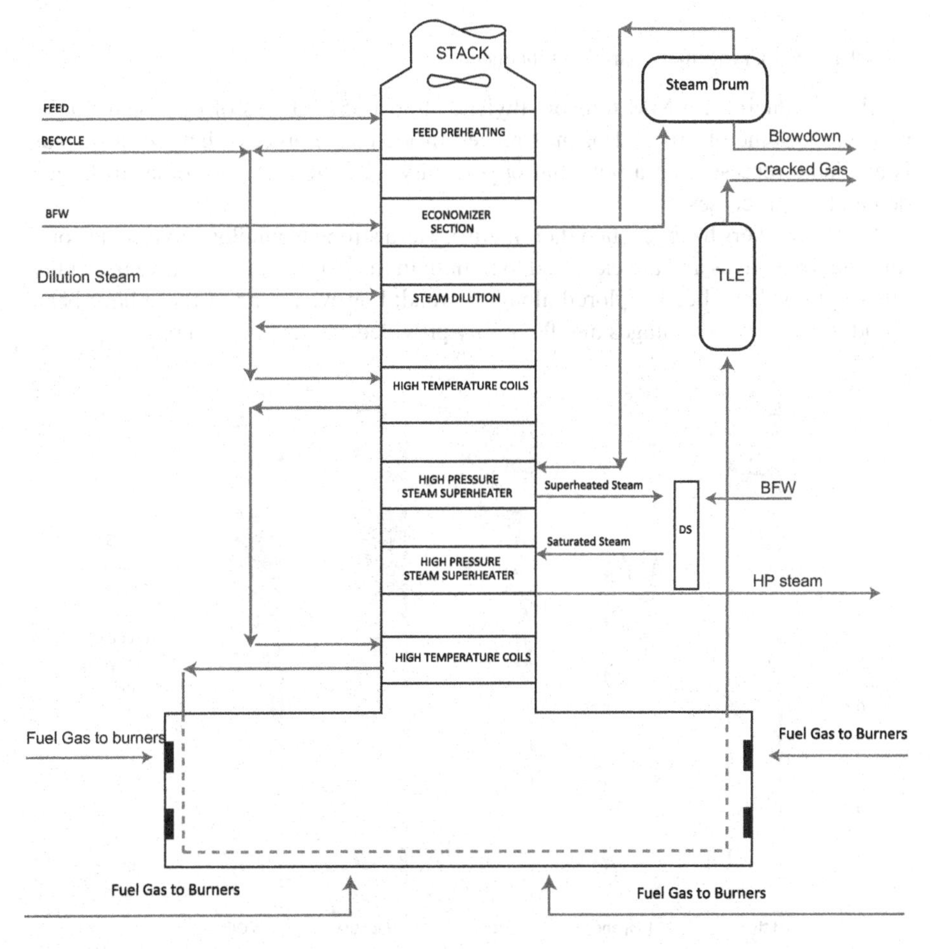

FIGURE 2.4 Typical pyrolysis furnace configuration

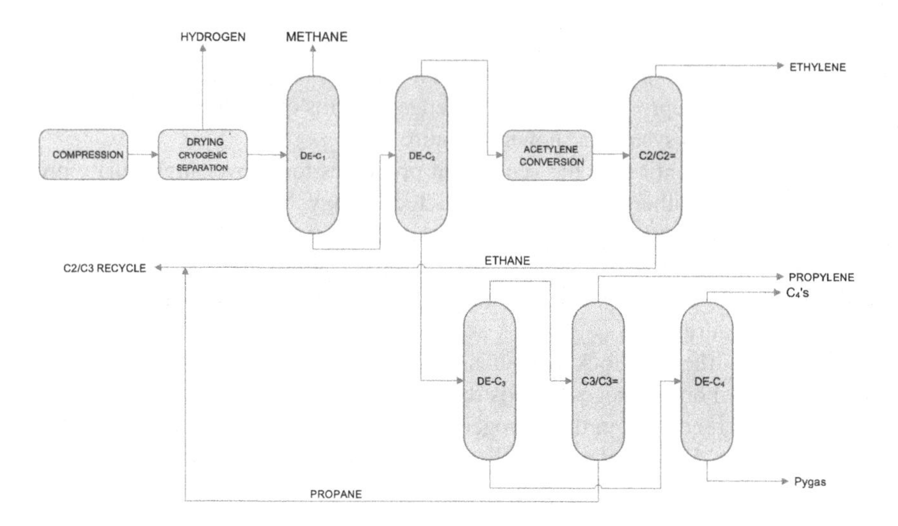

FIGURE 2.5 Cryogenic separation sequence.

cracker producing 1.5 MM tons of ethylene almost 95,000 tons of hydrogen can be recovered. Some of this hydrogen is reused inside of the process, but still a surplus is available representing a potential opportunity for synergies with other hydrogen demanding processes.

Steam crackers have reached their maturity as a strong technology to produce olefins and derivatives and are expected to remain in the market for several years. Many other options have been explored along the road; however, none of these have been found to have the advantages and flexibility provided by steam crackers.

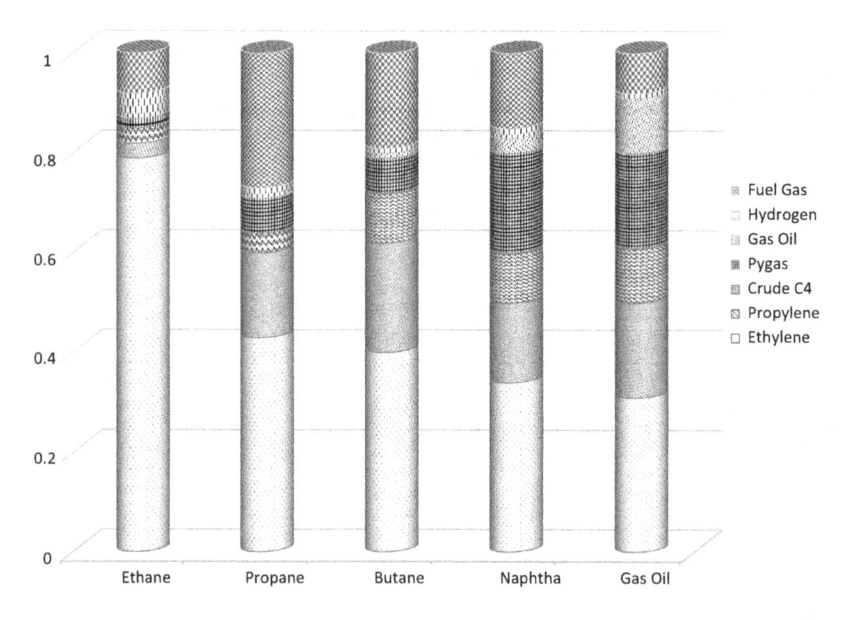

FIGURE 2.6 Typical cracker yields per feedstock.

Among the most relevant attempts are the coal gasification which makes use of the Fischer–Tropsch process to produce heavy olefins that can be cracked to obtain ethylene. This technology has not been competitive enough to displace the steam crackers, but coal-rich areas (e.g., China) are still looking for ways to improve these kinds of routes to take advantage of their local resources.

Another alternative route still in the radar is the direct cracking of crude oil, and the main limitation is the excessive amount of coke formation in the coils which has been tried to mitigate by either an improved vaporization of the crude oil, by using modified catalytic processes, or a combination of both. A brief sketch is presented in Figure 2.7 to provide a generic representation of this hybrid concept. The approach behind this development is to integrate the crude distillation process with the cracking of the stream to produce the maximum amount of light hydrocarbon components useful for production of olefins. The integration starts when the crude oil is preheated using the convection section of the cracking furnace. From this stage, all the light components stripped from the crude are directed to the cracking section, while the remaining liquid flows in countercurrent with steam to enhance their vaporization and recover the maximum amount of light components, and the stripped liquid can still be cracked over a catalytic bed which at some point needs regeneration, but hydrogen is provided to reduce the accumulation of coke. Similar alternatives have been explored; however, the quality of crude oil and the required profile of components (ethylene vs. propylene vs. aromatics) demand a very flexible design with capabilities to handle a wide range of crudes (not only by density but by level of contaminants). Further developments of this approach may become feasible as the demand of fuels gets reduced, and that is why this technology path will remain on the radar.

2.2.3 Aromatics Production

Similar to the situation of ethylene, benzene, and other related aromatic products (toluene, xylene) were originally produced from coke ovens, but in the case of the aromatics, not too many technologies evolved to replace the coke as main aromatics feedstock for several years, since aromatics demand grew at a different pace.

Substantial amount of aromatics were produced from coke ovens, but most of them were employed to supply the recently born fuels market, since many of the aromatic components represented a good resource to improve the octane rating in the fuels. However, the remaining volumes of aromatics used for chemicals had a great impact on the development of the chemical industry; one important example is

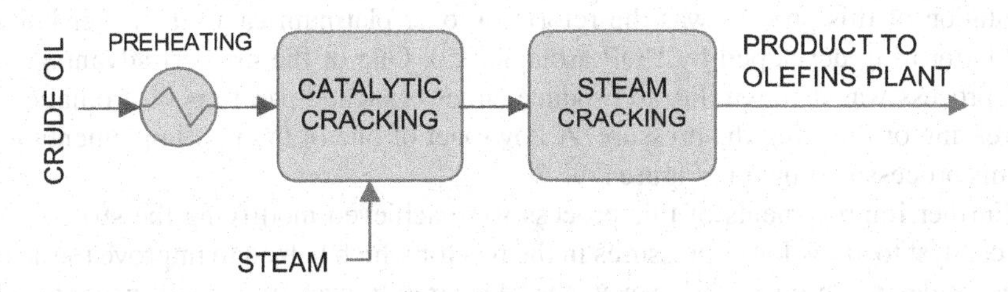

FIGURE 2.7 Cracking of crude oil for olefins production.

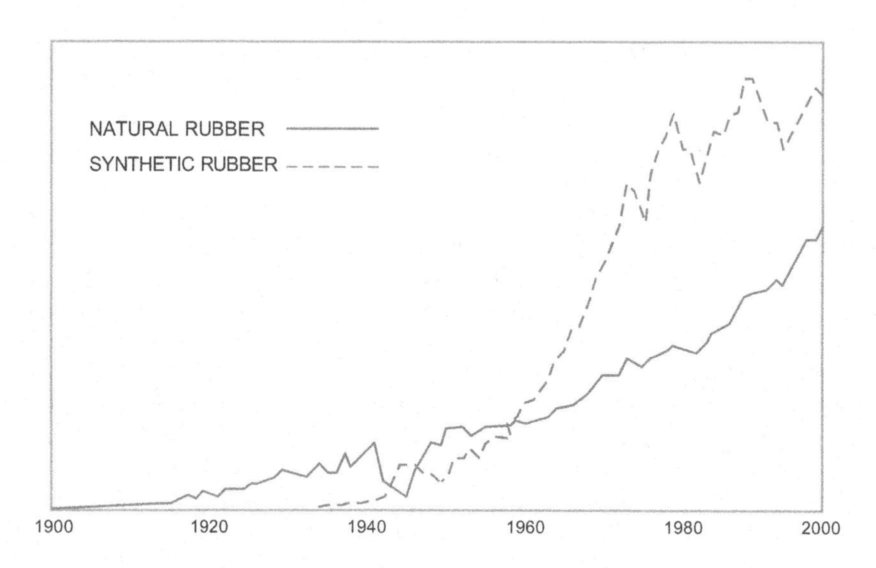

FIGURE 2.8 Production of natural vs. synthetic rubber.

the replacement of natural rubbers with synthetic derivatives such as nitrile rubbers which apparently provided a better resistance than the natural materials. Figure 2.8 illustrates the evolution of feedstocks to produce rubber.

Once the demand for synthetic rubber started to climb, alternative supply of aromatics from crude oil was slowly introduced. Then, benzene from refineries contributed to fill the gaps in the demand, especially when new catalytic reformers replaced the thermal cracking process to increase the octane rating in gasolines.

Initial development of catalytic reforming consisted of dehydrogenation and dehydroisomerization of naphthenes toward aromatics at high temperatures, but catalysts had to be continuously regenerated to remove contamination from coke. Therefore, research efforts were focused on maximization of time that reactors would be kept on stream which was achieved by either continuous regeneration (on fluidized catalytic beds) or catalyst improvements.

Process based on fluidized beds was initially implemented in the USA with some deficiencies that were hard to overcome, and at some point, the efforts were focused mainly on fixed-bed reactors using noble metals working at higher pressures (~3500 kPa) to increase catalyst stability and increasing the temperature close to the maximum levels that the catalyst could withstand. A successful example of the implementation of this process was the reforming over platinum catalyst further known as Platforming developed by UOP around 1950. One of the greatest advantages of this process was its flexibility to produce either branched paraffins or aromatics by increasing or lowering the pressure. A flowsheet of one of the first implementations of this process is shown in Figure 2.9.

Further improvements of this process were achieved modifying the structure of the catalyst to allow lower pressures in the reactor which helped to improve the selectivity of desired products. However, the regeneration cycle times still needed to be optimized. Alternative technologies tried to solve the problem by using cycling or

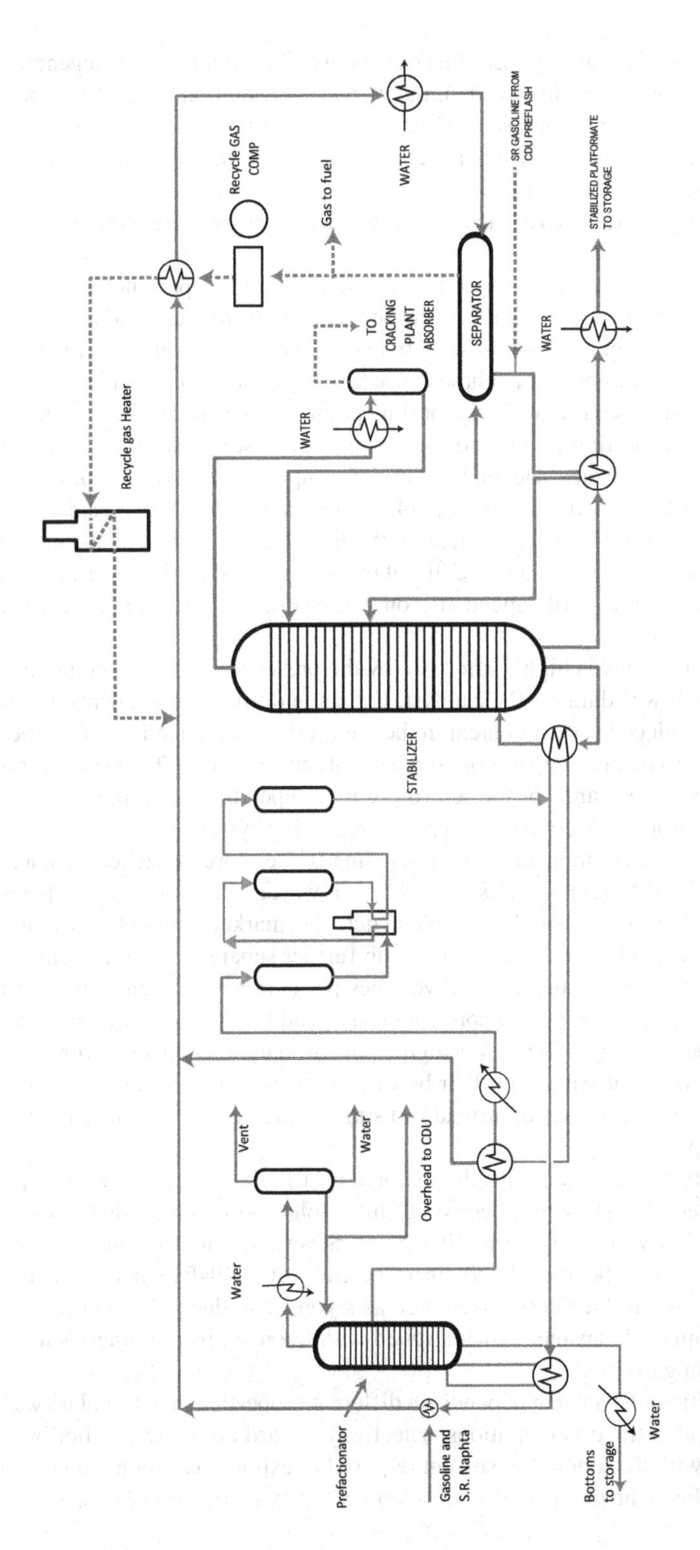

FIGURE 2.9 Reforming with fixed-bed reactors.

swing reactors in parallel which alternate between production and regeneration to keep a continuous operation of the units, while the evolution of the fixed-bed reactor landed in a new continuous catalytic regeneration process allowing the reactors to achieve very high selectivities to reformate, minimizing the conversion to light products at so much lower pressure.

Flowsheets of the different technologies developed over the time are shown in Figures 2.10–2.12.

For any of these technologies, the optimum operating parameters depend not only on the specific application (octane improving, aromatics production, yield to a specific component, etc.) but also on the composition and source of the feedstock (light or heavy straight run naphthas, cracked naphthas, coker naphthas, etc.); for example, feedstocks with low initial boiling point have negative effect on the octane, while high final boiling point feedstocks are more susceptible to coke contamination over catalyst surface. Depending on these inputs, the whole set of operational variables can be optimized, for example, temperature, pressure, residence times, catalyst selection or hydrogen content on the feedstock or intermediate reactors; all aligned to reach the most adequate distribution of aromatics. However, proper tuning of reforming units will depend also on the selection of performance of upstream and downstream units.

Catalyst efficiency is highly impacted by the presence of specific components (or poisons) which will damage the catalyst. For some of these components, the inhibition effect is reduced if their content in the feed stream is reduced, but for others, the reactor needs to be out of operation to regenerate the catalyst. Therefore, in this case the adequate performance of the reforming unit depends also on the quality of the feed and the proper selection of the upstream treating system.

Main aromatic components from reforming process are benzene, toluene, and a mixture of the different isomers of xylene. However, the ratio among themselves is very variable and normally doesn't match the market demand distribution for each component. Therefore, after reforming further separation or interconversion is needed to supply the actual demand volumes per component. Hence, the optimization of the reforming unit is not complete if these additional units are not included.

One of the challenges to recover high-value aromatic components from the reformate stream is the proximity of their boiling points with non-aromatic components, as shown in Table 2.1; hence, instead of using simple distillation, alternative routes had to be explored.

One solution was found with the extraction of high-value aromatic components using an external agent with selective affinity (solubility) with aromatics over other components but with a higher boiling point. Several agents (solvents) and process configurations have been explored since the end of the 1940s when the demand of benzene rose to produce derivatives such as styrene (produced from ethyl benzene) or nylon (from cyclohexane) which aligned with the need from refineries to remove benzene from gasoline.

The selection of a solvent depends on different properties such as polarity, chemical stability at extraction conditions, selectivity toward aromatics, or boiling point; these along with the amount of solvent required to extract maximum content of aromatics have been the parameters to develop several extraction processes. Some of the

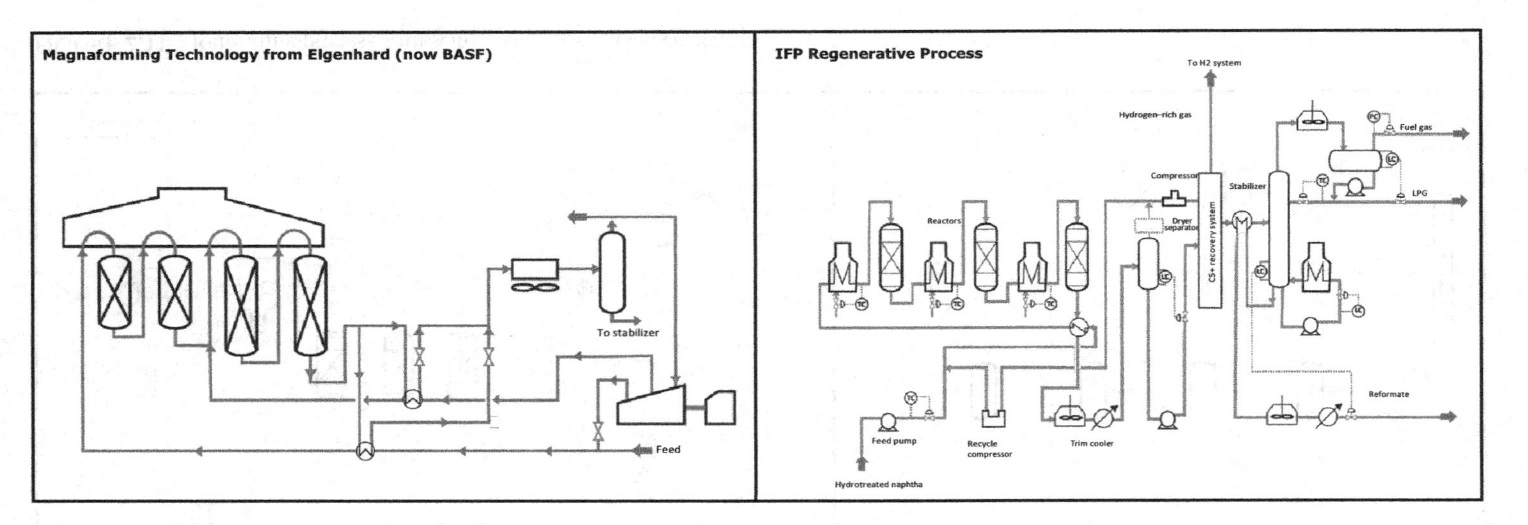

FIGURE 2.10 Reforming processes with intermittent catalyst regeneration.

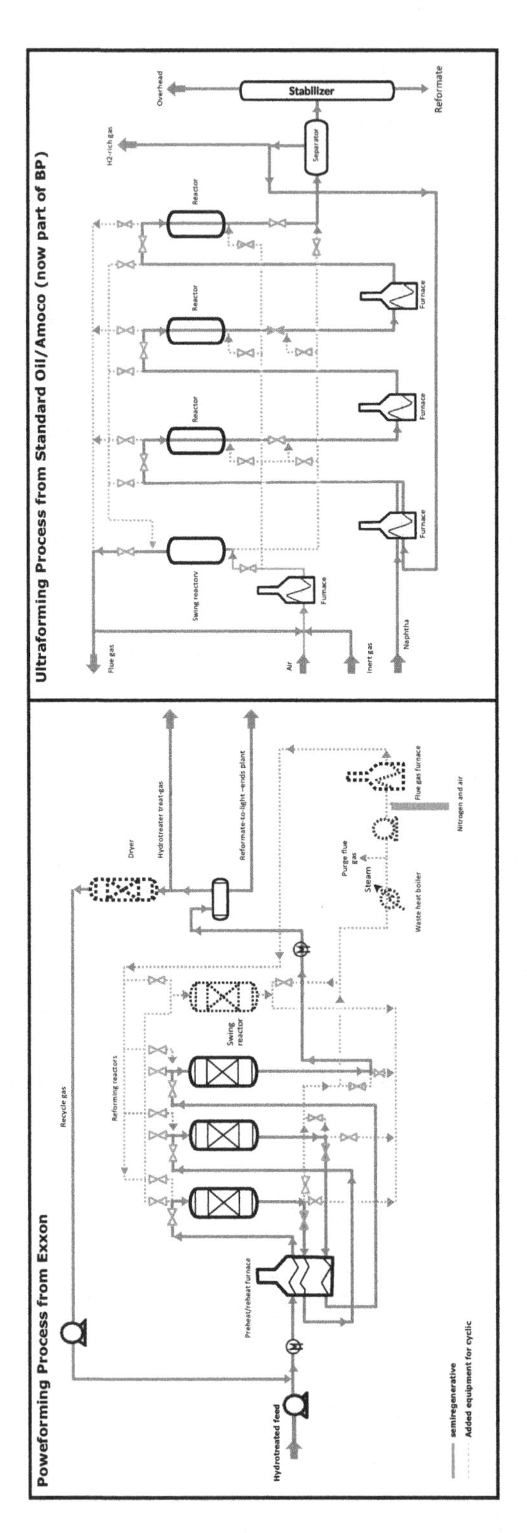

FIGURE 2.11 Reforming process with semi-regenerative fixed catalyst bed.

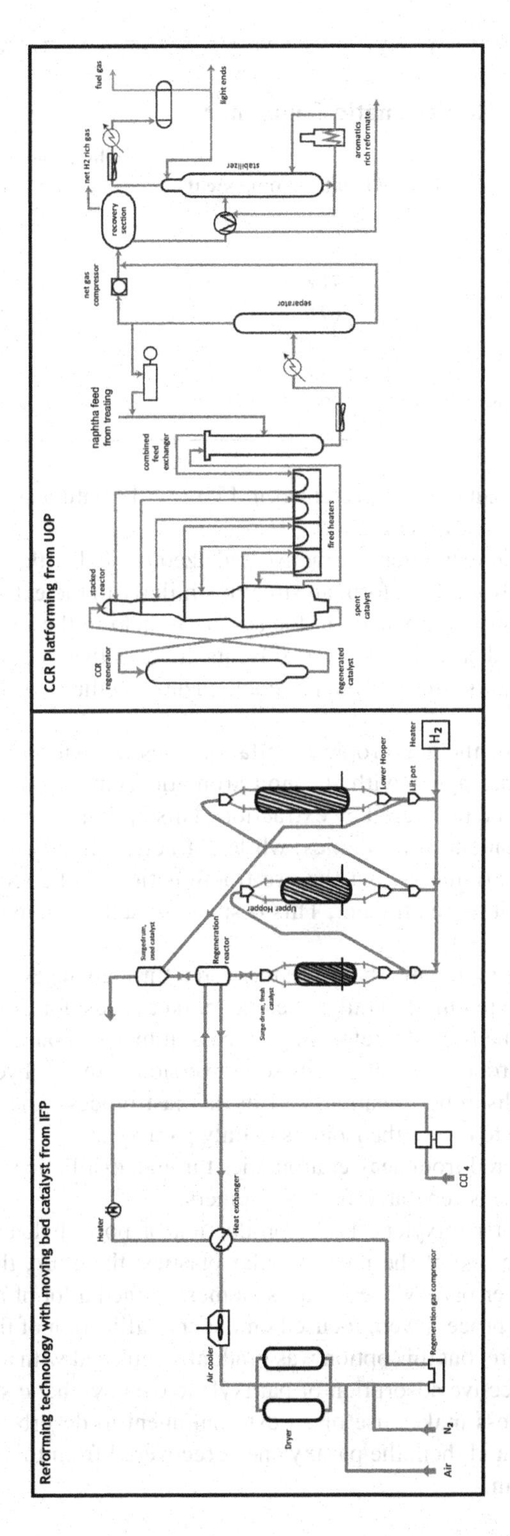

FIGURE 2.12 Reforming process with continuous regeneration of catalyst bed.

TABLE 2.1
Boiling Point of Non-aromatic Components

Component	Boiling Point of Component	Boiling Point of Azeotrope with Benzene
Benzene	80.1	
Cyclohexane	80.6	77.7
Methylcyclopentane	71.8	71.5
Hexane	69.0	68.5
2,2-Dimethylpentane	79.1	75.9
n-Heptane	98.4	80.1
2,2,4-Trimethylpentane	99.2	80.1

most representative solutions are presented in Figure 2.13 with a summary of their operational parameters in Table 2.2.

In addition to liquid extraction, extractive and azeotropic distillation can be used to recover aromatics from the reformate stream; similarly to the extraction process, an external component is introduced in the system to help in the separation. In the case of the extractive distillation, similar solvents are effective, but their purpose is to change the vapor pressure of aromatic components, facilitating the stripping of non-aromatics.

On the other hand, the azeotropic distillation uses an external component to produce an azeotropic agent with the non-aromatic component, and then, the azeotropic agent can be recovered by extraction. This option is feasible mainly for streams with high content of aromatics, while extractive distillation can be used with lower levels of aromatics, but the pre-fractionation of the feed stream may be required to make it more efficient, This is specially adequate to split benzene–toluene mixtures.

Once the mixture of aromatics has been recovered through the extraction process, further operations are required to tailor specific market needs; for example, depending on yields of toluene from the reforming process, it may be converted to benzene or xylenes. Possible routes to achieve these conversions are the hydro-alkylation, trans-alkylation, or disproportionation, and in this last process, the disproportionation can be selective to one of the isomers mainly para-xylene.

Xylenes are recovered from heavier aromatics through distillation, then if needed high purity paraxylene is separated from its isomers.

One of the uses of paraxylene is the production of polyethylene terephthalate (PET) which became one of the most popular plastics; therefore, the research on technologies to recover paraxylene from its isomers gained a lot of attention. First developments of this process were focused on the crystallization of the desired isomer at low temperature, but this option was gradually replaced with a more efficient process based on selective adsorption of paraxylene over synthetic structures such as zeolite. This process makes use of an external agent to desorb the paraxylene from such structure, and then, the paraxylene is recovered from this agent through a fractionation column.

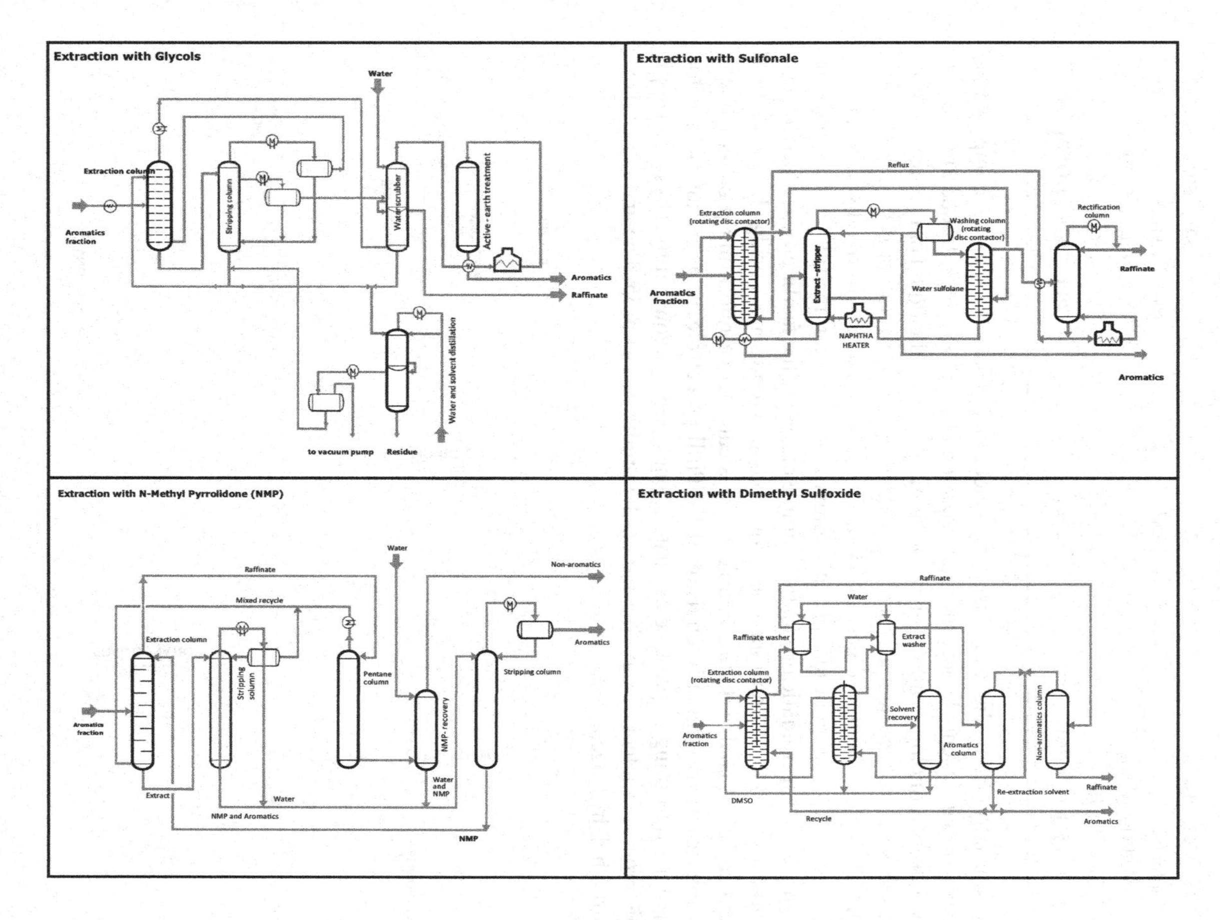

FIGURE 2.13 Liquid extraction process to recover aromatics with different solvents.

TABLE 2.2
Process Parameters for Aromatic Extraction Processes

Solvent	Boiling Point of Solvent, °C	Extraction Conditions
Diethylene glycol	245	130°C–150°C, 500–800 kPa
Sulfolane	287	100°C, 200 kPa
N-Methylpyrrolidone	206	20°C–40°C, 100 kPa
Dimethyl sulfoxide	189	20°C–30°C, 100 kPa
N-Formylmorpholine	244	180°C–200°C, 100 kPa

Normally from the three isomers, meta-xylene has less value compared to ortho- or paraxylene; hence, the isomerization of meta-xylene is frequently needed. Available technologies allow for isomerization in either the liquid or the vapor phase, and selection of the adequate technology depends on the overall targets of the facility considering that liquid isomerization has lower conversion rates of ethylbenzene.

A typical configuration of an aromatics processing facility is presented in Figure 2.14. An aromatic production complex is one of the facilities with the strongest connections to the oil refining operations, but also represents a bridge between other petrochemical chains. One example is the steam active reforming process (shown in Figure 2.15), originally envisioned by Phillips Petroleum as an alternative for catalytic reforming to produce aromatics, but nowadays (owned by ThyssenKrupp), this process is a technology focused on the production of propylene and isobutylene through dehydrogenation.

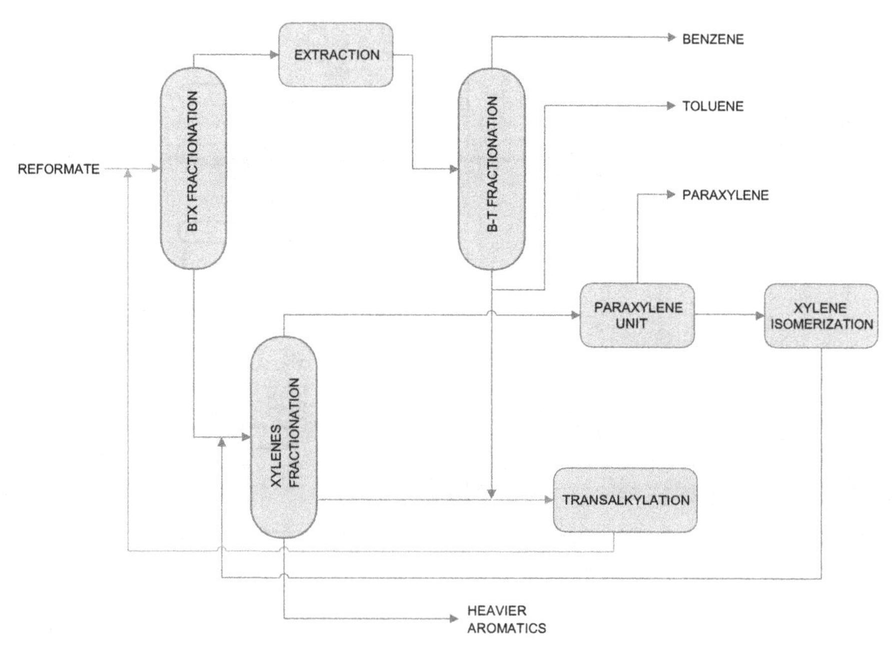

FIGURE 2.14 Typical configuration of a BTX processing unit.

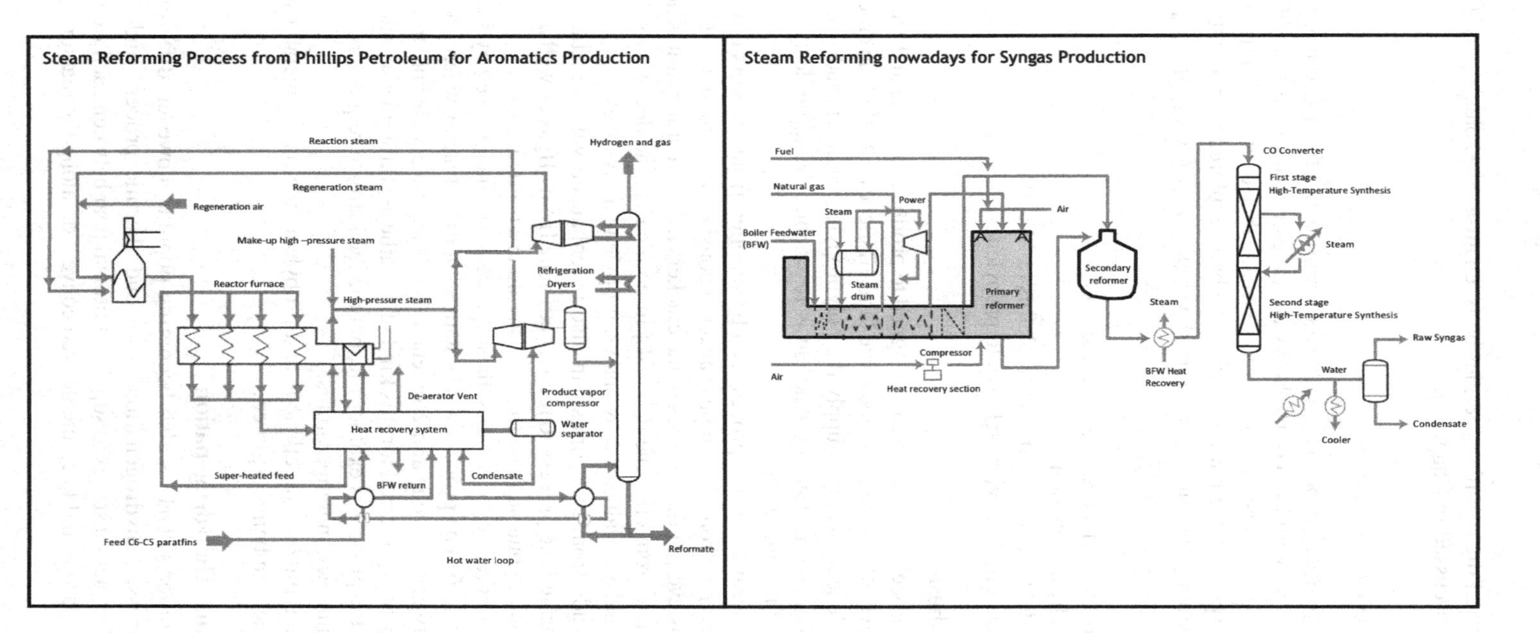

FIGURE 2.15 Evolution of steam active reforming.

Further details of the specific synergies between aromatics, olefins, and fuels production will be discussed in Chapter 4.

2.2.4 Propylene Production

Historically, propylene has been available as a by-product from two main sources, ethylene crackers fed with LPG or naphthas and fluid cracking catalytic (FCC) units in the refineries. But it was not until recent years that dedicated units to obtain propylene on purpose as the main product have been explored as alternative supply of this olefin. Among the most important developments for on-purpose propylene production are

- Metathesis of ethylene and heavier olefins
- Propane dehydrogenation
- Methanol dehydration (methanol to olefins or MTO)
- Interconversion of heavier olefins into propylene.

2.2.4.1 Metathesis

The metathesis process was originally designed to produce ethylene and butene from propylene; however, the increase in the scale (and efficiency) of ethylene crackers and the rise of propylene demand to supply the polypropylene market made this process unattractive. Since, the metathesis reaction is reversible, the same design has been readapted to produce propylene.

Typical feed streams for this process can be supplied from crackers or refineries. When the feed comes from a naphtha cracker, a considerable amount of C4's would be available. However, in the case of ethane crackers a dimerization unit is needed to produce the butene from ethylene (C2–C4); in both cases, the decision is driven by the market demand and value of ethylene vs. propylene derivatives.

Propylene yields from FCC units can also be enhanced with a metathesis unit; however, an intermediate process to recover ethylene from offgas as well as hydrogenation units to obtain butene from the C4's stream are needed.

A typical configuration for the metathesis process commercialized by Lummus is shown in Figure 2.16. The process works with a catalytic reactor where the metathesis reaction between ethylene and 2-butene takes place but also is able to isomerize 1-butene to 2-butene (preferred for this kind of metathesis) through a secondary catalyst. In general, the process is exothermic, but the feed stream requires preheating to achieve optimum reaction temperatures.

In order to maximize both, ethylene and propylene alternative metathesis routes using only butenes are being explored.

2.2.4.2 Propane Dehydrogenation

Propane dehydration (PDH) is also becoming popular; however, different to the metathesis process, dehydrogenation is an endothermic process and therefore requires multiple reaction stages to supply additional heat between each step to avoid high temperature drops and keep the desired conversion along the reactors.

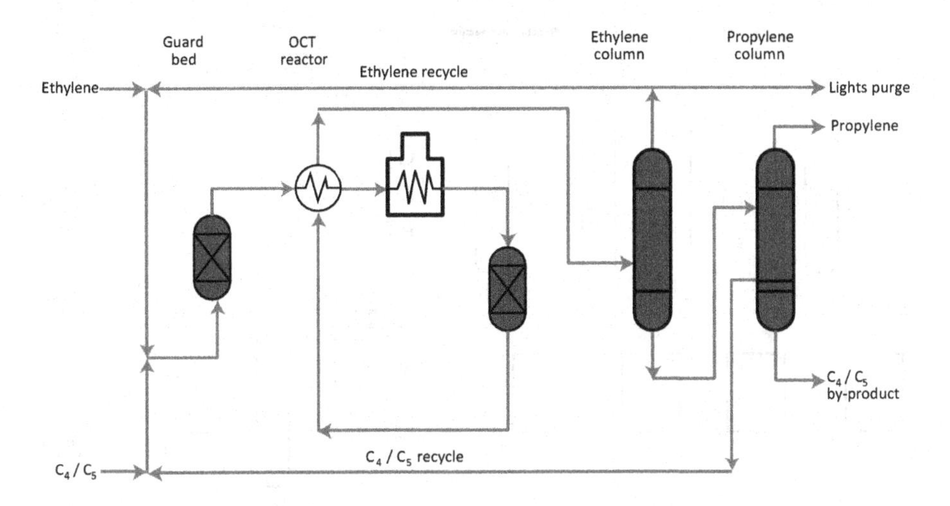

FIGURE 2.16 Production of propylene via metathesis.

High temperature not only favors the conversion of the feed but also increases the possibility of thermal cracking. Therefore, temperature should be carefully controlled to achieve the maximum selectivity toward propylene. The process works efficiently at low pressure, and then, reactors work just above atmospheric pressure to avoid oxygen leaking into the system (to avoid a flammable/explosive environment).

One of example of successful implementations of the PDH process is Oleflex from UOP (shown in Figure 2.17), which makes use of a similar catalyst and regeneration system from their Platforming process, complemented with a low-temperature recovery system to remove the propylene from the reactor effluent.

Similarly, to the catalytic reforming technologies, PDH has a version with cyclic regeneration (instead of continuous), and this technology is available from Lummus (shown in Figure 2.18) and works with three reactors in parallel, one in service, one in regeneration, and one in purge mode. In this version, hydrogen is not recycled, and high selectivity is achieved in the reactors working under vacuum.

2.2.4.3 Methanol to Olefins (MTO)

Another technology that has been becoming available in large scale during the last years is the production of olefins from methanol. The innovative side of this option is the fact of making an alternative source available to supply olefins especially for regions with naphtha crackers where the competitiveness is driven by oil prices, because this alternative, the methanol and therefore the olefins, can be sourced from natural gas, coal, biomass, or even carbon dioxide (yes carbon dioxide!).

Most of the MTO plants have been built in China where coal is highly available. Originally, the research was focused on delivering a process to produce gasoline from methanol, and this was the starting point to move toward the production of olefins. However, it was found that the catalyst involved in this process was not entirely selective to a specific olefin. Therefore, the lines of research moved to either the research of efficient schemes to recover a specific olefin from the reactor effluent or

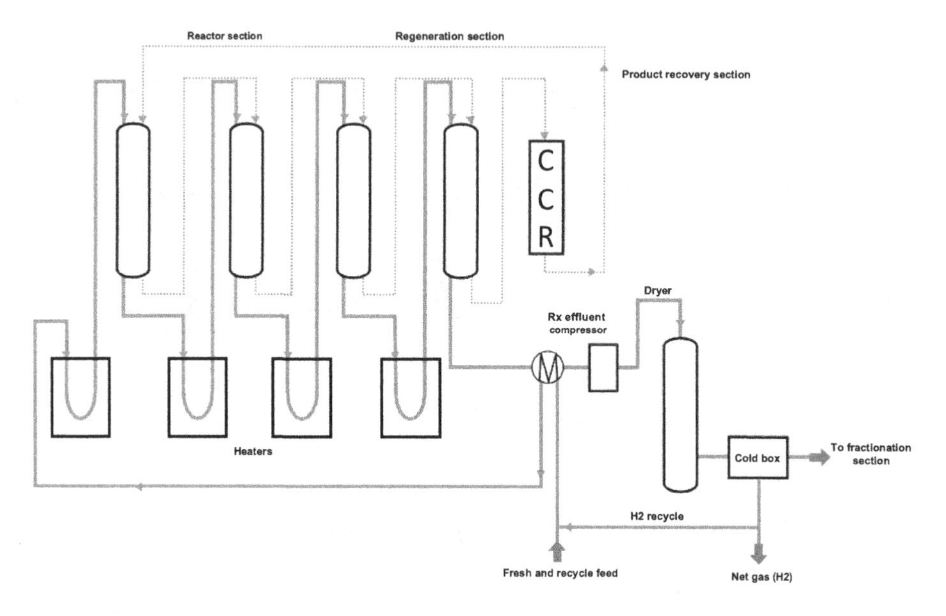

FIGURE 2.17 Production of propylene via propane dehydrogenation (UOP Oleflex process).

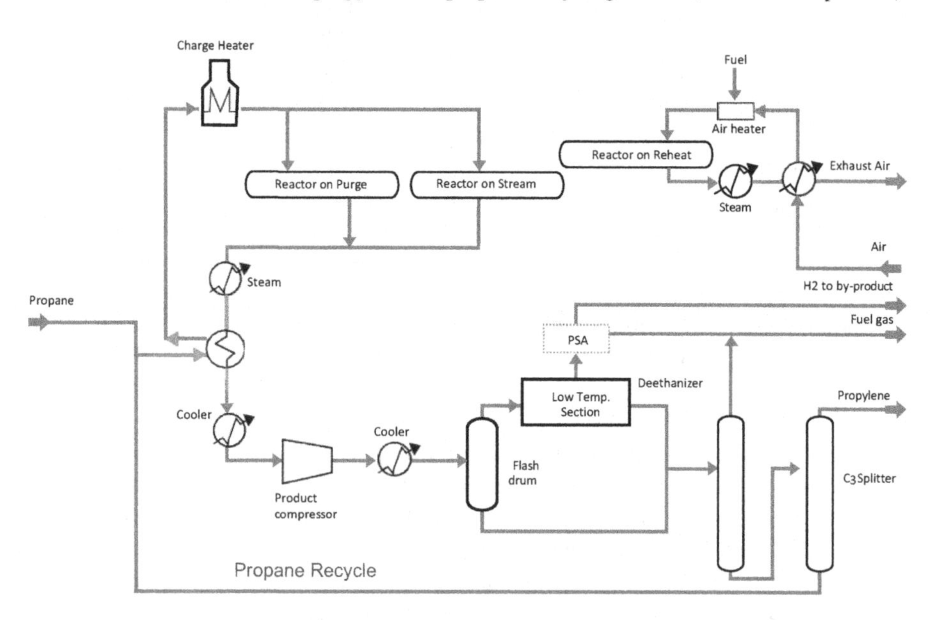

FIGURE 2.18 Production of propylene via propane dehydrogenation (Catofin process).

the development of catalysts with specific selectivities to olefins (in particular propylene) to avoid the need of complicated separation schemes.

Examples of these technologies are illustrated in Figure 2.19. The first flowsheet corresponds to the UOP/hydro process which is based on a catalyst with high selectivity to ethylene and propylene and uses a fluidized bed to regenerate the catalyst, with

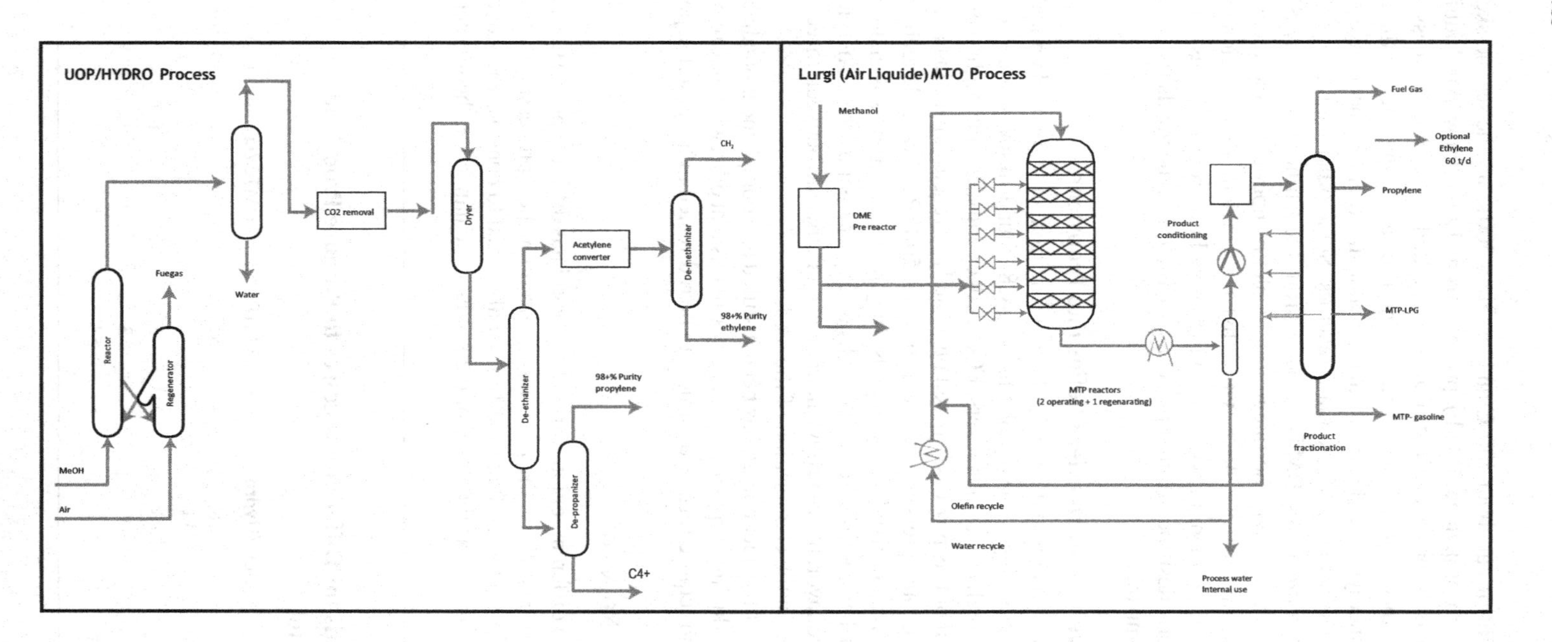

FIGURE 2.19 MTO technologies.

typical reaction conditions at about 500°C and less than 500 kPa. The second sketch shows the methanol-to-propylene (MTP) process from Lurgi (now Air Liquide) which produces not only propylene but also gasoline and LPG. This process has fixed-bed reactors with semi-cyclic regeneration and uses a preconverter as the first step is to produce dimethyl ether (DME) from methanol, and then, further conversion from DME to olefins is achieved in the main reactors. Some other technologies are in development, for example, the DMTO process from the Chinese company SYN Energy Technology which needs the integration of olefins upgrading and recovery processes such as metathesis of C4/C5 with ethylene, ethylene dimerization, and olefins fractionation all of them provided with the know-how from Lummus; this is the option with the broadest spectrum of products. In Table 2.3, the yields for each of these options are compared.

2.2.4.4 Interconversion of Heavier Olefins into Propylene

An additional option to increase the yields of propylene from other petrochemical or refining units is the cracking of heavy olefins (C4–C8) found in low-value streams from upstream units such as steam crackers, FCCs, coker units, etc. into light olefins. Typical configuration of this process is based on a catalytic reactor operating at high temperatures (~500°C) and pressures between 100 and 500 kPa. Available technologies have either fluidized or fixed catalytic beds. In general, the process is endothermic but requires further cooling; hence, some heat recovery is possible. The yields from this process are favorable to propylene rather than ethylene between three and five times depending on the composition of the feed.

The development of this technology has been justified as a complement of adjacent units to add value producing propylene from heavy streams that otherwise may remain as naphtha that would require further treatment to be redirected to the gasoline pool.

2.2.5 Petroleum Refining

Crude oil available in different regions around the world has different properties even for the same location. Therefore, the facilities required for their processing not only have to be tailored for such specific properties but also need to consider the value of the mixture by itself as well as the value of the products demanded in the market.

TABLE 2.3
Product Yields for Different Processes to Produce Propylene from Methanol

	UOP/Hydro (%)	Lurgi MTP (%)	DMTO/Lummus (%)
Propylene	18	31	17
Ethylene	14	-	17
Gasoline	3	8	1
LPG	6	3	3
Other components	59	58	62

The key properties to define the value of a specific mixture are the density and sulfur content. These properties define the level of complexity in a refinery, while a light mixture will be easy to split into different components requiring lower energy consumption, and the level of sulfur will define how complicated are the facilities required for its treatment. In Figure 2.20, representative crudes from different locations are plotted as a function of these two properties.

Based on these parameters, it is easy to understand that heavier crudes will have less value and therefore may represent an opportunity as an alternate feedstock to produce petrochemicals, increasing the value of the crude by converting it into products with higher margins.

Refineries can be as simple as a topping unit (mainly distillation), or as complex as a deep conversion unit (with several units to improve yields of products with high value and to reduce levels of sulfur). The integration of additional operations to convert available streams into products with higher value comes with an impact on the level of complexity in the refinery and therefore a consequent impact on the energy consumption.

For more complex units, the energetic efficiency of an integrated scheme will play a critical role in their implementation; for example, if an existent refinery runs with old and inefficient facilities, the integration or revamp of a new unit with dual purpose (i.e., a new or improved reformer) may contribute with hydrogen in excess, an improved energetic consumption, and the maximization of yields toward aromatics components to replace the decrease of the demand on the gasolines pool. But on the other hand, if an existing unit is already efficient and the profits from their products are substantially high, the integration with other facilities will impact its performance and will be feasible only if the total integrated scheme has a higher added value. Figure 2.21 shows the relationship between energy consumption with the level of complexity of a refinery and how an integration can improve (or reduce) the efficiency of the operations.

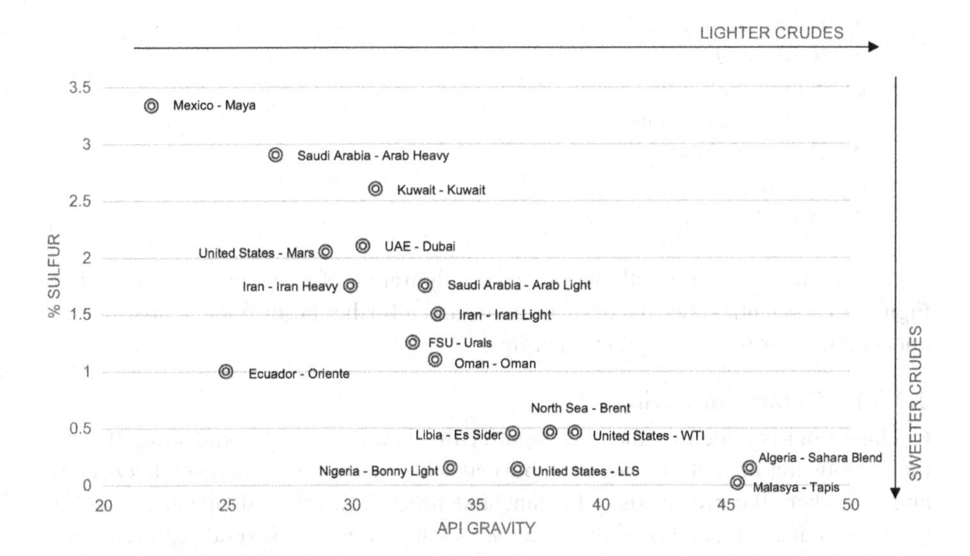

FIGURE 2.20 Properties of representative crude oils.

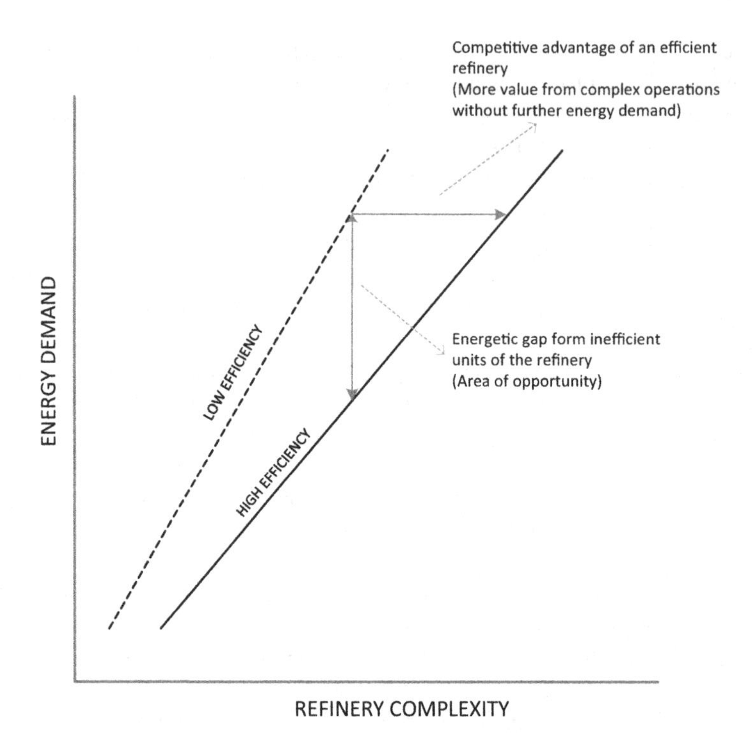

FIGURE 2.21 Impact of a new unit in the complexity and energetic consumption of a refinery.

The facilities with most of the opportunities for integration can be briefly classified as

- Octane Improving
  - Isomerization
  - Reforming.
- Upgrading
  - Catalytic cracking
  - Hydrocracking.
- Desulfurization
  - Hydrotreating.

A generic block diagram with the typical configuration of a refinery is presented in Figure 4.1 to identify the role of these units, and a further description is provided to understand their role in possible synergies.

2.2.5.1 Octane Improving

Octane rating is a measurement to quantify the detonation level (knocking) during the combustion of fuel; it's been established with an arbitrary reference between 0 and 100 where 0 corresponds to heptane and 100 to 2,2,4-trimethylpentane (one of the most branched isomers of the octane). Octane ratings for several representative hydrocarbons are shown in Figure 2.22, in which it is visible how olefins, branched paraffins, and aromatics have higher octane ratings compared to lighter paraffins.

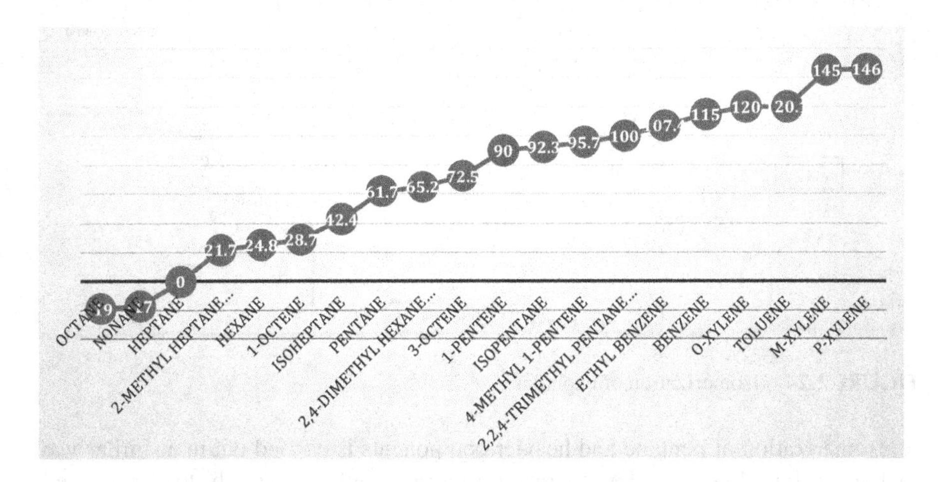

FIGURE 2.22 RON of different hydrocarbons.

2.2.5.2 Isomerization

One of the most common processes to increase octane rating of gasoline is the isomerization. This process was originally developed to produce isobutane which is further required to produce alkylate and MTBE, but typically is useful to increase the octane of the light naphtha with composition of C5–C6, since the linear paraffins are converted into their correspondent isomer which has a branched structure.

The isomerization of butanes and light naphthas is depicted in Figures 2.23 and 2.24. Both are based on a reaction system which is favored with low temperatures. However, low temperatures also reduce the rate of isomerization, and therefore, a catalyst is required. Isomerization is achieved through an equilibrium reaction, and therefore, the amount of the isomer in the feed will have a negative impact on the conversion, and this is specially particular for the isomerization of butane. Therefore, if high content of isobutane is expected, it may need to be removed upstream of the reactor.

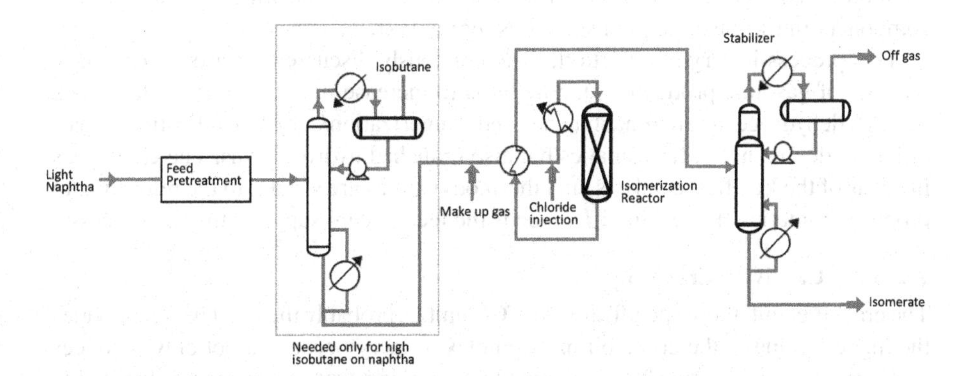

FIGURE 2.23 Isomerization of butanes.

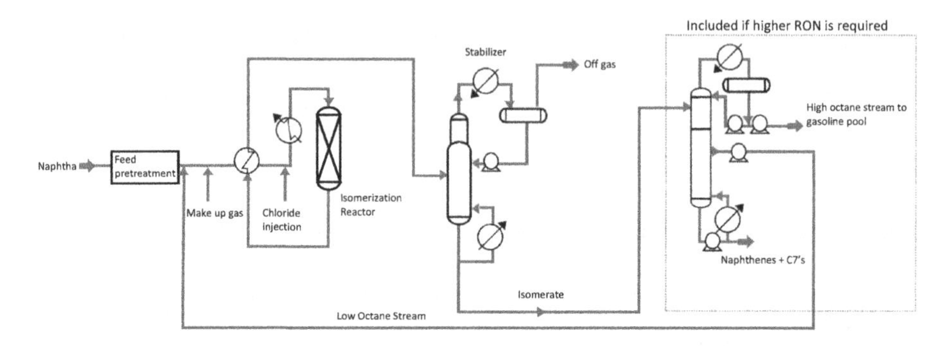

FIGURE 2.24 Isomerization of naphthas.

Isomerization of pentane and heavier components is carried out in a similar way as butane, but obviously, the kinetics are more complex, specially for C6 isomers where yields are split in components with different octane ratings; then, if the amount of low octane products is considerable, the isomerate can be fractionated to recover high octane streams and recirculate low octane components back to the reactor.

Isomerization is one of the main users of hydrogen in the refinery mainly for two reasons: the activity of the catalyst is affected by the presence of sulfur; therefore, hydrotreating may be required depending on the sulfur levels of the naphtha. On the other hand, hydrogen is required inside the process not only to inhibit the production of olefins and coke in the reactor, but also to produce cyclohexane from the saturation of benzene when this component is present in the feed stream. This is of great help on the refineries especially for those with strict restriction in benzene content in the gasolines; however, two valuable resources to establish an integrated scheme with petrochemicals are being depleted: hydrogen and benzene.

2.2.5.3 Gasoline Reforming

While the octane rating of light naphthas is improved through the isomerization process, reformation is a better fit to increase the octane of heavy naphthas. If light naphthas are fed to the reformers, these will be cracked (rather than reformed), and on another hand C6 components will reformate into benzene which is not a desired component in the gasoline pool since it is highly toxic.

This process is very similar to the one previously discussed for aromatics, but in the case of gasoline production, the target is to increase the octane of the feedstock through dehydrogenation of naphthenes and isomerization/dehydrocyclization of paraffins producing not only aromatics but also branched hydrocarbons. The challenges in terms of the kinetics, catalysts, and thermodynamics are similar to the reforming to produce aromatics. The main difference is the feed stream supplied to each process.

2.2.5.4 Catalytic Cracking

The cracking unit, more specifically the FCC unit, is probably the plant which provides the highest value to the crude oil in a complex refinery because it not only produces high gasoline yields from heavy components coming from the vacuum distillation unit, but also produces a specific rating of octane in the gasoline, olefins, and LPG.

The conception of this unit goes to the 1930s where the first catalytic units using fixed-bed reactors came on stream. These units were commonly identified as the Houdry crackers referring to Eugene Houdry, a French engineer who set the first stones of the catalytic cracking process. Houdry's initial work was focused on obtaining gasoline from lignite and oil derived from the mineral using a catalyst based on clay. His work was of interest to American companies that were looking for sources of high octane gasoline demanded by the automobile industry. By 1937, the first Houdry unit was already in service processing 15,000 barrels per day of crude with yields to high octane gasoline reaching 50%, and the process ran at high temperature (~400°C) using a molten salt heating system which was also used to recover the heat available in the flue gases resulting from the catalyst regeneration to produce steam; his technology required a cyclic regeneration of the catalyst and the controls were based on automatic timers.

Few years after, pushed by the need of supplying aviation fuel demanded to support the fleets in the WW2, improved versions of the catalytic cracking process were available incorporating a fluidized bed for regeneration, and from that point up to present days, the catalytic cracking has been one of the most important units in the refining process.

In general, FCC units include a main reactor working continuously in synchrony with a regenerator connected with transfer lines to move the catalyst between them, and the hydraulic design of this configuration was challenging because the required gas flow had to provide the minimum fluidization velocity to overcome the pressure drop through the catalytic bed and keep a uniform fluidized homogeneous pattern where the mixture gas–catalyst can move as a single homogeneous phase. The regeneration of the catalyst is achieved by combustion of the coke accumulated on the catalyst with air. Therefore, another challenge was to ensure a proper isolation of the hydrocarbons in the reactor from the regeneration circuit to avoid a flammable and explosive atmosphere.

The FCC reactors evolved rapidly with different configurations to overcome several issues, but from early configurations, it was found that some degree of cracking was taking place in the transfer line with regenerated catalyst to the reactor. Hence, further developments of FCC technology strive for longer lines (risers) to enhance the cracking through this section in a more diluted phase reaching better yields and selectivity to higher octane gasoline.

Latest configurations of reactors include not only a cyclone at the top of the riser for an efficient catalyst–gas separation to remove the product and avoid further undesired reactions, but also enhanced top riser devices to ensure proper seal of hydrocarbons. New designs aim for shorter residence times to improve the selectivity of cracking.

On the other side, the spent catalyst flows downwards in countercurrent with steam to remove some of the remaining hydrocarbons and then is fed to the regenerator where the coke is removed. The developments of regenerator systems have been focused on achieving a complete combustion of coke not only to reduce the amount of CO emissions but also to improve the quality of regeneration; in addition to a complete combustion, additional heat on the catalyst is available and can be removed producing steam. The remaining energy is still used to vaporize the fresh feed once this gets in contact with the regenerated catalyst. Figure 2.25 shows a typical configuration of an FCC reactor with the regeneration system.

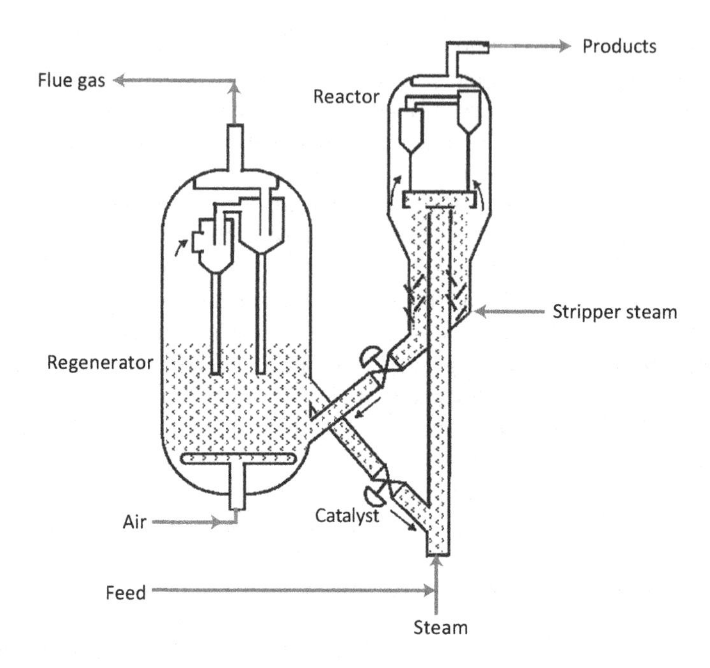

FIGURE 2.25 FCC reaction/regeneration unit.

Another line of research on the FCC units is the catalyst; first, FCC units adopted the clay (silica/alumina) catalyst used by Houdry, but in the case of FCC, the goal is to achieve not only the efficiency of the catalyst but also the adequate fluidization, lacking on this kind of material. Several improvements have been incorporated in the structure of catalyst, and some of them make use of zeolitic materials which have shown a better suit for the demands of the FCC units. These configurations along with injection of additives provide the FCC units with great capabilities in terms of flexibility to handle different feed compositions as well as a diversified range of products.

The typical feedstock of an FCC unit is the gasoil produced from vacuum distillation, which is a mixture with different levels of paraffins, olefins, naphthenes, and aromatics. All of these components are converted into lighter olefins in an FCC; however, the chemistry and kinetics are not as simple as that. Additional side reactions take place in the reactorsuch as the isomerization of olefins and naphthenes, dehydration of naphthenes and aromatics, transalkylation of aromatics, etc. The extent of each reaction depends on the specific conditions of the reactor and of the catalyst selected; therefore, the design can be tailored to achieve yields of any specific component, light olefins, LPG, gasoline, or even aromatics. A traditional FCC unit can be operated at high severity to increase the amount of light olefins, but still a considerable amount of naphtha with high octane ratings will remain in the reactor effluent. In order to promote the conversion of this naphtha into higher levels of aromatics and olefins, more selective catalysts need to be incorporated. The yields of light olefins can increase from 10% on conventional process up to 40% with modified catalyst with selectivity to petrochemicals.

The flexibility of the FCC units is visible not only in the product yields but also in the feedstocks. Typical design is based on vacuum gasoil, but heavier streams can also be cracked operating at higher severities (i.e., higher temperatures). However, the associated price is an increase in the coking rates. Therefore, additional considerations come into place such as specific residence times, better injection of the feed at the entrance of the riser to enhance its vaporization, and improved regeneration systems (i.e., multistage regeneration).

2.2.5.5 Hydrocracking

The hydrocracking, similar to the catalytic or thermal cracking, was implemented to upgrade the value of heavy fractions of crude oil by producing fuels. Early developments of this technology commenced in the 1930s. However, the catalytic process with operation pressures much lower than hydrocracking draws more attention, and most of the efforts were focused on that direction. But once the reforming process became more popular, a considerable amount of hydrogen as by-product was available, making hydrocracking a more feasible option to supplement FCC units.

The main advantage of hydrocracking units is their flexibility; while the efficiency of an FCC is decreased with lighter feedstocks, a hydrocracking unit can handle any stream from vacuum distillation as well as lighter cuts coming from the atmospheric distillation to produce different products depending on the feed which can go from distillates or naphtha, jet fuel, or even LPG. Therefore, this technology is normally implemented to fill specific gaps in the optimization of refining operations.

The configuration of the process sequence depends on the purpose of the unit, and it can go from a simple unit for partial conversion or removal of contaminants to a more complex circuit with staged reactors and recalculations to achieve a more complete conversion of the feed. A typical configuration for different sequences of hydrocrackers is presented in Figure 2.26.

2.2.5.6 Hydrotreating

Hydrotreating is a common technique in refining to remove undesired components from different streams. Typical contaminants are sulfur, nitrogen, chloride, peroxides, phenol, olefins, polynuclear aromatics, organometallic components, etc. Final purpose is not only to avoid contamination of the catalyst in downstream units but also to meet specific product specifications especially from environmental regulations. Hence, hydrotreating is considered a mature technology that has evolved in parallel with other catalytic process, but in the past decades, the environmental restrictions have pushed for more efficient parameters in the design of these units.

The chemistry and kinetics of these processes are very well known, and having specific catalyst for different applications, process conditions depend on the type of feedstock, and both temperature and pressure requirements increase with the molecular weight of the stream; for example, naphtha hydrotreaters work at ~300°C and 1 MPa, while the treatment of heavy residue requires temperatures close to 400°C and pressures above 15 MPa. In the same way, hydrogen consumption for heavy streams like the residue can be 25 times of the amount required by naphtha. The typical configuration of a hydrotreater is presented in Figure 2.27.

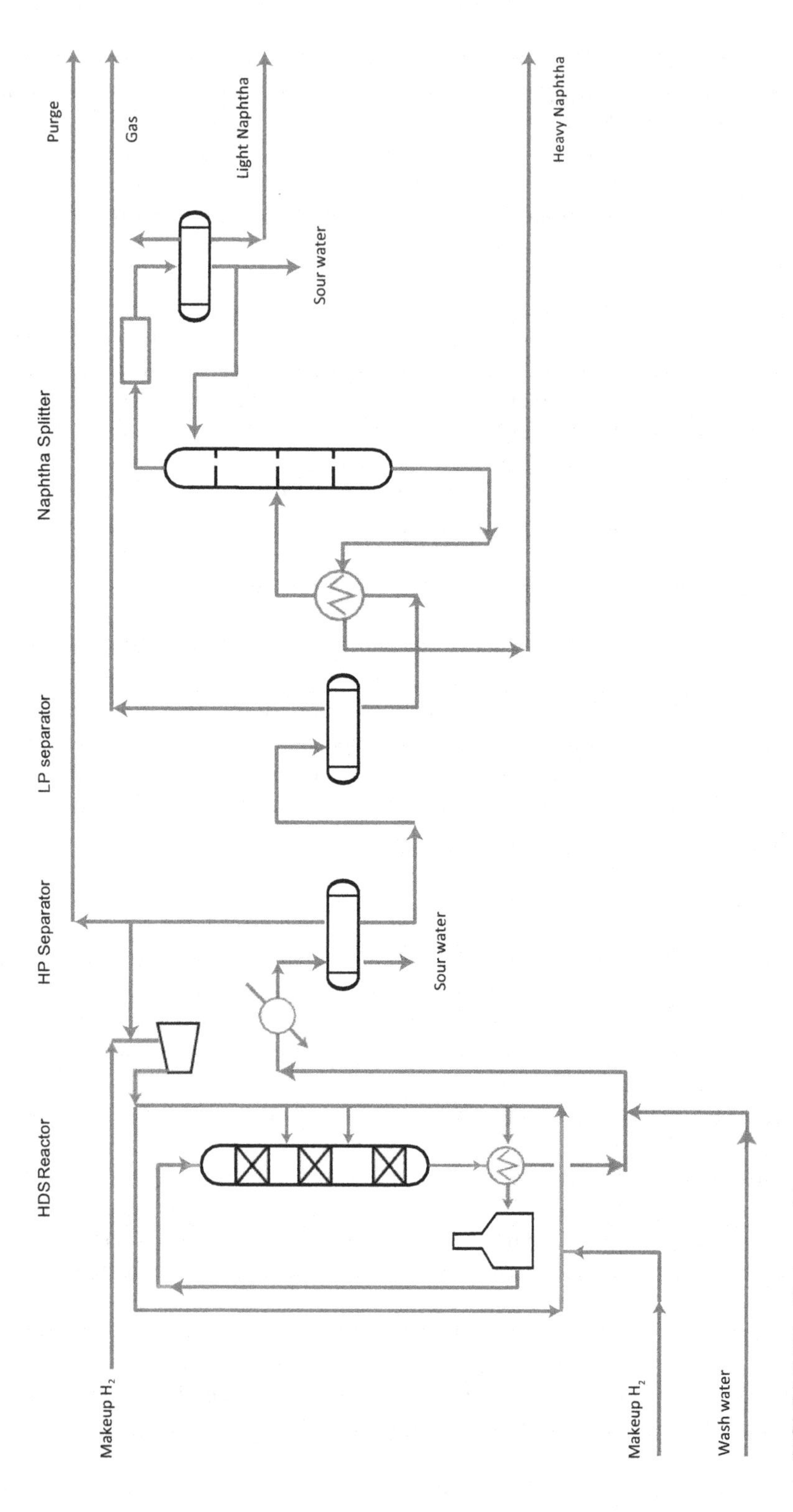

FIGURE 2.26 Hydrocracking process.

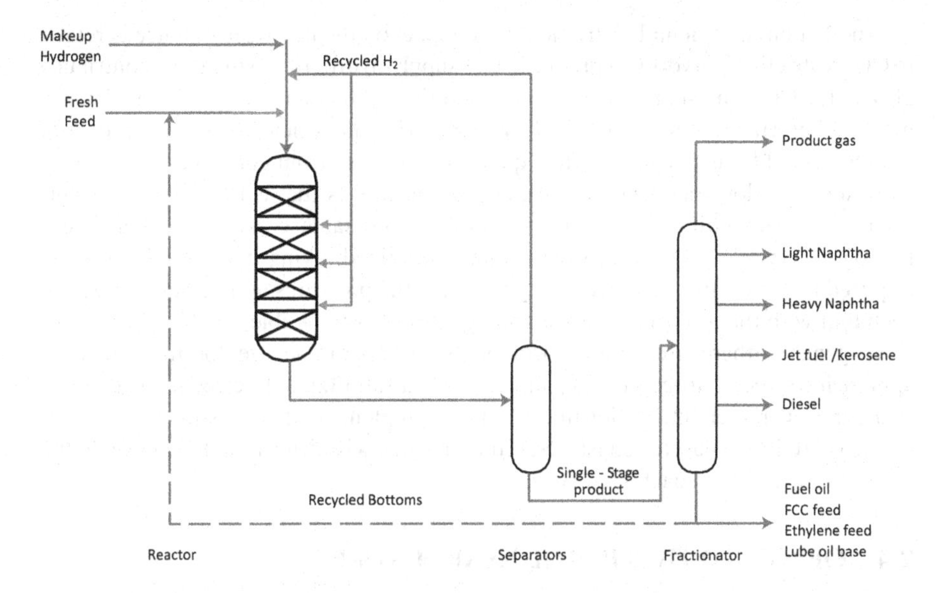

FIGURE 2.27 Hydrotreating process.

Alternative routes to remove contaminants, such as sulfur, have been explored, for example, biodesulfurization, pervaporation with membranes, extraction, and oxidation. However, the hydrotreaters are still a predominant technology to remove sulfur and other contaminants, and therefore are still a main consumer of hydrogen in the refineries.

2.3 ROLE OF HYDROGEN IN SYNERGIES

Refineries are one of the main consumers of hydrogen through the hydroprocessing units described above. Global hydrogen consumption from refineries was estimated at 38 Mt/y[1] by 2019. While fossil fuels demand may not grow at a high pace, environmental regulations (i.e. levels of sulfur in fuels) are expected to be more stringent requiring additional sources of hydrogen. Today, only approximately 30% of the hydrogen required in the refineries is self-sourced mainly from reforming units, and almost 40% of the needs is supplied with steam-methane reformers. Very reduced amount of hydrogen is provided from integrated petrochemical facilities. On the other hand, the cost of hydrogen production is estimated between 1 and 2 USD/bbl, equivalent to 15%–40% of the margins of refining. Hence, every opportunity to reduce the demand of external hydrogen supply can be quantified with a positive impact on the refining margins.

It is also important to consider that in the long term, the requirement of hydrogen will be impacted by the replacement of fossil fuels with alternative options (i.e., electric cars); therefore, incremental addition of on-purpose hydrogen units may not be justified.

[1] The Future of Hydrogen, report prepared by IEA for the G20 Japan, June 2019.

Another environmental restriction with impact on the hydrogen balance is related to CO_2 restrictions. Hydrogen production to supply hydroprocessing units contributes 20% of the CO_2 emissions from refining operations. Therefore, the addition of steam-methane reformers will be inhibited by more stringent policies on carbon reduction or at least will have a negative effect quantified in costs from penalties.

In terms of logistics, another element to consider is the fact that the units supplying and demanding hydrogen don't work at the same conditions. Therefore, a hydrogen network has to be redesigned and optimized taking into consideration the required infrastructure to move the hydrogen at the pressures and purity required by each unit with the minimum capital and operating cost possible.

In general, there are enough arguments to support a case for the integrated approach for the hydrogen management between refining and petrochemical operations. However, such integration may need to be implemented on existing or revamped units. Therefore, a careful constructability analysis will determine how feasible this kind of integration in each facility is.

2.4 OPPORTUNITIES IN THE NEAR FUTURE

2.4.1 Reduction of Fuel Consumption

Historically, the demand of fuels has been driven by the energetic consumption for transportation, and this will still be supplied mainly by fossil fuels in the following 20–30 years, especially for non-OECD countries.[2] Therefore, the availability of streams from crude oil to use as a feedstock for petrochemicals will depend on an efficient management in terms of energy of these streams in the refining systems.

However, there is an interesting opportunity brought by the introduction of alternative fuels which will have a negative impact on the demand of diesel and gasoline, and can be accentuated after the unexpected reduction on the consumption of these fuels caused by the restrictions imposed to avoid the spread of the COVID-19 in early 2020.

These fuels could be exported to countries with increasing demand gasoline or diesel, but depending on the extent of the impact from COVID-19, their production can be better replaced with products with higher value for the petrochemical market, to increase overall profitability levels.

2.4.2 Strategies to Share Infrastructure

The integration of product chains is not restricted to a simple optimization of assets in a specific geographic location, especially for global organizations with facilities in strategic locations around the globe. The capital and operational cost of a new processing unit for a petrochemical product is considerably high compared to the cost of transportation of the same product from another facility with surplus capacity available. Nowadays, the logistics to move products between different locations are very efficient and represent a great aid to create more robust and optimum integrated schemes.

[2] According to 2019 reports from the EIA.

Today after the movement restrictions resulting from the COVID-19, it is understandable that these alternatives will be highly challenged, but proper mitigations can be explored like the increase on operational stock of products to find global synergies.

2.4.3 Environmental Challenges

One of the drivers in the petrochemical industry has been the demand for plastics. The contribution from the plastic sector has been fundamental for the evolution of several economic sectors, medicine, education, communications, food processing, etc. However, this evolution had a price, and now after several decades of development, a huge environmental concern will impact these trends of prosperity for the plastics and therefore the petrochemical industry.

Globally, a huge amount of trash is being generated from single-use plastics, and materials that are used by the final consumer for a short amount of time and therefore with a high frequency. Today it is not clear how the governments will define policies to control it, but, for example, countries such as Chine have started to ban the import of plastic waste for recycling due to the impact on air pollution from the recycling facilities.

Regardless of the pattern of future regulations, it is wise to consider this element as a part of the synergies of the petrochemical industries with refining since both will still depend on the fossil fuels which at some point will face more criticism once alternative energy and material sources found feasible ways to become available to compete with the oil and gas industry.

BIBLIOGRAPHY

American Chemical Society; The Houdry Process for Catalytic Cracking; April 1996.

Franck H.G.; Stadelhofer J.W.; *Industrial Aromatic Chemistry*; Springer-Verlag, Berlin, Germany; 1988.

Hydrocarbon Processing; *Petrochemical Processes Handbook*; Gulf Publishing Company LLC., Houston, TX; 2018.

International Energy Agency (IEA); The Future of Hydrogen, Report prepared by IEA for the G20 Japan; June 2019.

Kniel L.; Winter O.; Stork K.; *Ethylene Key Stone to the Petrochemical Industry*; Marcel Dekker Inc. New York, NY; 1980.

Little D.; *Catalytic Reforming*; PennWell Publishing Co. Tulsa, OK; 1985.

Meyers R.; *Handbook of Petrochemicals Production Processes*; McGraw-Hill, USA; 2019.

Meyers R.; *Handbook of Petroleum Refining Processes*; McGraw-Hill, USA; 2016.

U.S. Energy Information Administration; International Energy Outlook 2019 with projections to 2050; September 2019.

3 Economic and Market Outlook

Before we provide the technical assessment of the synergies between oil refining and petrochemicals, we need to have a good look at the global and regional economic framework in which strategic investment decisions are being made. A good understanding of markets, economic drivers, and outlooks is essential to any discussion of the refining and petrochemical markets. In fact, due to their interaction with and dependence on the global energy market, we will give an overview of the trends and predictions for global energy supply and the impact these trends have on the oil refining and petrochemical sectors. The data we show in this book is based on our own assessment of energy supply and demand, and takes into consideration the global energy outlook reports of all major energy companies that post their assessment reports on the public domain. It is very difficult to predict the future behavior of markets and industries, so the statements made in this chapter reflect our opinion and may differ from the opinion of other industry experts and authorities.

As an example, none of the outlooks that were published a few years ago would have predicted the COVID-19 pandemic and its devastating impact on the global economy, on consumer behavior, and on oil prices. While we most certainly need to consider the short-term effects of COVID-19, we also have to assume that in the not too distant future, the global economy will recover and consumers as well as manufacturers and providers will reach a new equilibrium that will determine if the current predictions will prevail or if adjustments to the global energy outlook will be required. The reduced economic activity as a result of mandatory restrictions has caused changes in the energy supply and demand patterns. It is our opinion that it will be necessary to adjust the methodologies and metrics that we use to monitor the energy markets and to predict future trends. There will be lessons learned and newly discovered opportunities to increase efficiencies in workflows and processes. There will be changes on consumer behavior based on the experience from living with the current restrictions caused by COVID-19. However, we also believe strongly that the overall trends and predicted developments will continue to be valid. Changes will most likely have impact on regional markets and specific market sectors. There will be a decrease in employer commuting traffic to and from work in high population density areas as companies and employees got more comfortable with the benefits of remote, home office work. Long-distance travel will stay at a lower level than it was before for the same reason. Remote working and video conferencing gain in acceptance, and people learn how to use it more efficiently and confidently. These are just two examples of change that will occur. However, the global trends and the envisioned energy transformation as predicted will continue, and the statements derived from those predictions will stand despite COVID-19 and its global impact.

3.1 GLOBAL ENERGY MARKETS

The global energy market is dominated by fossil sources and will be for a long time. Despite all efforts to drive toward a transition away from fossil fuels and energy sources, we must be realistic in what we can expect and what can be achieved in what timeframe. We will explain in more detail our view on the transition and what can be achieved by when in this chapter. For the purpose of simplicity, we look at three fossil energy sources and define the non-fossil energy sources in three groups. This makes a total of six energy sources to look at

- Coal
- Crude oil (conventional and unconventional)
- Natural gas (conventional and shale gas)
- Renewables (biofuels, bioenergy, hydropower)
- Nuclear (traditional and small-scale)
- Wind/Solar (onshore, offshore).

Table 3.1 shows our assessment of the current global market share for each of these sources, for the predicted market share in the year 2040, and for the short-term and long-term trends we believe these groups will experience.

In general, the overarching theme for developing the global energy and power generation sector is decarbonization. This will be measured against the decrease in global emissions of carbon dioxide, which has been determined to be one of the main culprits for causing global warming and changes in the global climate. Since the combustion of all fossil fuels results in the formation and emission of carbon dioxide, all energy companies have agreed to work on concepts and technologies that will allow the industry to reduce carbon dioxide emissions, by decarbonizing the sources for energy and power generation. This transition will be executed in two phases, of which only the first phase is clearly defined.

- Phase 1: Transition from sources with high carbon intensity to sources with low carbon intensity. This phase is defined by the commitment to phase out coal and crude oil, the two fossil energy sources with the highest carbon intensities,

TABLE 3.1
Global Market Shares Power Generation

		Global Power Generation		
Source	**2020 (%)**	**2040**	**Short-term Trend**	**Long-term Trend**
Coal 27%	19	Strong decline	Strong decline	
Crude oil	31	32%	Slight gain	Stagnant, slight decline
Natural gas	23	26%	Strong gain	Strong gain
Renewables	12	13%	Stagnant	Slight gain
Nuclear	5	6%	Slight gain	Slight gain
Wind/Solar	2	4%	Strong gain	Strong gain

from power generation and from manufacturing of fuels. At the same time the industry will increase the use of natural gas, the fossil fuel with the lowest carbon intensity, for power generation and fuel manufacturing. In parallel, a lot of the capital investment will go into the development and implementation of other fuel and power sources as renewables, wind and solar.
- Phase 2: Transition from low carbon intensity sources to non-fossil energy sources. And here is where the predictions get a bit fuzzy. While there are technical feasible and commercially available alternatives, none of them has the potential to replace fossil energy sources yet.

The transition to low or zero-carbon-intensity energy sources and the resulting reduction in carbon dioxide emissions is not driven by the energy market, but by environmental concerns and political intervention. This fact complicates the analysis of the energy markets and explains some of the challenges we face in implementing the required changes.

The political journey to where we are today started in 1992 with the United Nations Convention on Climate Change (UNFCCC). During this convention, all participating parties or nations signed an agreement that states (1) that global warming is occurring and (2) that it is predominantly caused by human-made carbon dioxide emissions. As a countermeasure, it was agreed to reduce the emission of greenhouse gases (GHG), including carbon dioxide as the main component.

In 1997, the agreement from the UNFCCC was adopted in the Kyoto Protocol, a treaty developed by over 190 member nations. The Kyoto Protocol was entered into force in 2005. The first commitment period started in 2008 and ended in 2012. The second commitment period was developed under the name Doha Amendment in 2012. However, as of Q1 2020 only 137 member nations have signed the Doha Amendment. It requires 144 nations to accept the treaty to enter into force. But several nations have changed their approach to fighting global warming, one of them being the USA. These nations believe that a global approach to the problem does put certain nations into economic disadvantage and that a national, focused approach, defined by setting their own targets and by acting independently from other nations, is a more efficient and quicker way to fight global warming.

Enforcing a politically driven program bears certain risks that will cause severe resistance in the industry who must implement it. We have no intention to comment on global warming nor do we want to take part in the dispute about the interpretation of climate data and the conclusions drawn from that interpretation. But we want to highlight that enforcing the energy transition program gets its most resistance from two main sources:

- The lack of practicality and feasibility of utilizing certain energy sources due to availability and/or sustainability.
- The lack of economic benefit of utilizing certain energy sources, mainly driven by the cost of utilization.

Saving the planet is a noble cause and deserves all the support we can give. Ruining the global economy will not support that cause, but in fact cause more damage than it will do any good.

This sets the stage for the evaluation and discussion of all six energy sources and their future role in the global energy and power generation mix.

3.1.1 Coal

Coal is a solid fossil fuel that is formed from peat, which itself is formed from remains of plants in tropical wetlands that existed millions of years ago. The pressure of layers of rocks and soil that laid down on top of the peat transformed the peat into the rock formation that we know as coal. There are different kinds of coal with slight differences in composition and use:

- Lignite – also called brown coal; lignite is mostly used as solid fuel for electric power generation
- Sub-bituminous coal – like lignite, sub-bituminous coal is used as solid fuel for electric power generation, but is also a source for light aromatic hydrocarbons for the chemical industry
- Bituminous coal – bituminous coal is mostly black in color, but can be found in dark brown color as well; it's a dense, but soft rock that breaks easily and burns quickly; bituminous coal is used as solid fuel in power plants, as heating medium in manufacturing processes, for blacksmithing and as source to make coke for the steel industry
- Anthracite – a hard, black rock that burns longer and is used mainly in residential and commercial space heating

Other materials that would fall into this group are

- Peat – the precursor of coal; peat is mostly used in very specific application such as whiskey distilleries
- Graphite – natural graphite is used in electrodes for the steel industry as well as in batteries; powdered graphite is also used in pencils or as lubricant; it is very difficult to burn and, therefore, not suitable as solid fuel.
- Charcoal – charcoal can be produced by heating wood in an airless atmosphere; charcoal is produced mostly for private use as solid fuel for grilling.

The main component of coal is carbon. The composition of coal is completed by hydrogen, sulfur, oxygen, and nitrogen. The combustion of coal in the process of power generation or heating generates significant amounts of carbon dioxide as well as soot, mercury, and carbon monoxide. The air pollution caused by the decades of burning coal is main contributor to carbon dioxide emissions. It is estimated that up to 50% of the global carbon dioxide emissions stem from burning of coal. It also contributed to diseases such as asthma and cancer. In addition, the mining of coal in all its forms – deep underground mining, surface mining, or mountaintop mining – created severe environmental problems. All these factors are playing a role in the decision of many countries to abandon coal-based power generation and heating with a less carbon-intensive or zero-carbon fuel or energy source.

For example, if the carbon dioxide emissions are the baseline representing 100%, switching to other fuels or energy sources can lower the carbon dioxide emissions significantly. Switching to natural gas as fuel for power generation is the most common decision. We estimate that using natural gas can reduce the carbon dioxide emission by about 60%–40% of the base level set by coal. Some sources claim that the reduction can be as high as 80% to about 20% of the baseline emissions. Another example we would like to mention is hydroelectric power. We will talk about types of hydroelectric power plants later in this chapter. Hydroelectric power has been reported to reduce carbon dioxide emissions by approximately 99% based on typical carbon dioxide emission numbers reported for such facilities. However, there are reports that indicate that carbon dioxide and methane emissions from the decomposition of plants (trees, bushes, grass etc.) and organic matter that was or will be flooded by the artificially created water reservoir have been underestimated. These reports also highlight that the extremely low numbers of carbon dioxide emissions from hydroelectric power neglect the emission created during the construction and decommissioning of the hydroelectric power plants. However, even the corrected numbers still represent a huge reduction of >95% of carbon dioxide emissions compared to coal.

Despite all these negative effects of using coal as energy source, it is still heavily used in China (largest coal producer in the world), in South and Southeast Asia, and even in the USA. A lot of countries such as Germany have declared that they will shut down their coal-fired power plants within the next few years. Even China has announced that it will level out the use of coal for power generation. Only still growing economies such as South and Southeast Asia have plans to build and operate new coal-fired power plants. The main driver for this decision is the low cost of coal and its application in power generation as well as its availability. And developing countries may not put as much focus on issues such as pollution and health concerns as fully developed countries do. But the pressure of decarbonization will also reach developing countries, and it seems inevitable that coal as an energy source will decline over the next 20–30 years and will be replaced by natural gas and renewable energy sources.

There are only very few factors that could change that picture. Better, more efficient and less expensive gasification technologies could allow the use of coal and similar products such as petroleum coke in the generation of methane-rich gas that will be the feedstock for fuel manufacturing, petrochemicals, chemicals, and a source of energy for power generation. The commercial development of a direct digestion process for coal could also change its fate as direct digestion has no carbon dioxide or other emissions. Unless a technology like the ones mentioned above emerges and becomes commercially available and proves to be environmentally and economically acceptable, the fate of coal seems to be sealed.

3.1.2 Renewables

The group of renewable energy covers a broad range of materials and energy sources that – as the name says – can renew itself within a certain period. The advantage of this concept is that as long as the source renews, there will be no shortage. As a contrast, fossil fuels are being mined and produced from this planet and cannot be replaced or renewed. The amount of fossil energy sources we have available is limited. However, we still have a lot of coal, oil, and natural gas resources identified that are feasible to produce that we don't have to expect a shortage any time soon.

Renewables are often mistaken to be the same as biofuels or bioenergy. While these sources are part of the renewables group, there are other sources that are accounted for here as well.

The most common group in renewables is any fast-growing plant such as elephant grass or sorghum. These plants would be harvested and converted to ethanol, and regrown to create a cyclic process that would allow the conversion of these resources to fuels and energy. Some other plants that were considered or used for ethanol production are now protected as they are part of the food supply chain for humans. Examples are sweet corn and agave. Another group would be wood chips or wood pellets from the timber industry. There may be environmentally acceptable scenarios for growing trees as renewable energy source. But in many cases, the wood chips or pellets will be a by-product of preparing the wood for the construction, paper, and furniture industries. Another source that has been used in ethanol production is the residues from processing sugar cane. This is mainly attractive in countries that have a large sugar cane processing industry, for example, Brazil. The residue from harvesting sugar cane with machines and the wet bagasse from crushing the sugar cane are both good feedstocks for biofuels processing as they contain about 50% sugar and 50% water.

Another group of fast-growing, renewable energy sources is algae. Many companies do research on the growth of certain species of algae and its conversion to fuels or energy. In other cases, the focus is more on using algae as the basis for environmentally friendly fertilizer. Algae contain fats and sugar, which makes them a great source of energy. The fats can be converted straight to biodiesel, and the sugar can be converted to bioethanol. As algae can be grown in controlled environments and in dedicated plants, these facilities do not compete with the food supply chain for the animals in our seas and oceans. It still must be seen what yields of energy can be achieved on a sustainable basis and at what cost. The growth potential for algae as renewable energy source is difficult to predict until more data is available. However, we believe that algae will play a significant role in the future of renewable energies as this group will grow its market share continuously.

We also wanted to mention green or renewable hydrogen at this point. It is our opinion that hydrogen is not a primary energy source and it needs to be manufactured from other sources. Any hydrogen that is produced from methane (for example, in a steam-methane reformer) stemming from natural gas of the gasification of another fossil fuel must be considered energy from fossil fuels. Only hydrogen generated from renewable resources can be accounted for under this group of renewable energies. The challenges that still need to be resolved on a reliable and safe basis are the storage and distribution of hydrogen for use by consumers such as short range, private transportation or residential heating. Hydrogen will play a role, but as big of a role as some experts want to make us believe. The cost of manufacturing, storing, and distributing hydrogen in a safe and reliable manner will be limiting its growth potential and range of application.

Last, but not least, we also consider hydroelectric power as renewable energy. Via the global cycle of water (evaporation and rain), natural water reservoirs are refilled without use of artificial power (e.g., pumping). Therefore, hydropower is renewable. In hydroelectric power plants, the potential energy of water is converted into kinetic energy of water which can be used to turn the blades of a water turbine to generate electricity. The concept itself is very old. A lot of industries have settled next to rivers as they used the flowing water to turn a wheel which then acted as the driver for machineries such as big saws or grinders and mills. In our modern times, we know three types of hydroelectric power plants.

The most common type is the impoundment facility. This is basically the facility everybody pictures when talking about hydropower. It consists of a dam that is used to contain the water in a reservoir or pool. The water then is released in a controlled manner to flow downwards through huge pipes that at the end host the turbines and power generation portion of the plant.

The second type is the diversion facility. The abovementioned historic use of water as driver for machinery is an example for a diversion facility, as water from a river is redirected through a channel or multiple channels toward turbines driving generators.

The third type is called a pumped storage facility, and it is not a true hydroelectric power facility, nor can it necessarily be considered a renewable energy source. These facilities are used to store energy from another energy source by using that energy to pump water uphill into a pool or reservoir at higher elevation and then use it like described for an impoundment facility. The primary energy source to drive the water pumps can be solar energy, wind energy, a type of fossil energy, or nuclear energy. Depending on the type of primary energy source, these facilities can be accounted for as renewable energy or not. We also would like to highlight that the efficiency of using such facility is quite low due to the energy losses from pumping the water to higher elevation. However, in some cases it just might be the better alternative to the use of batteries or other storage technologies.

3.1.3 Wind/Solar

For many people, wind and solar energy are the solution for the future. We don't share this opinion, and we will explain in this chapter why these energy sources will play a role, but not a significant role in the global energy mix.

Wind energy is harnessed via installation of wind turbines either in single turbine applications for private or residential users (decentralized power generation) or in so-called wind farms that consist of multiple turbines. The number of turbines can range from 5 to 150 for a typical size. To show to what extreme this can go, the wind farm in Altamont Pass, California, USA, has more than 4800 turbines. In the past 5 years, a trend has developed for installation of wind farms offshore to minimize the negative environmental impact of wind farm installation. Onshore installations are considered harmful in a way that they do cause loss of habitat for wildlife and plants, and they support the fragmentation of the land. They also have been proven to increase the death rate for birds and bats. And leakage of lubricant oil from the turbines can cause minor oil spills. Looking at the complete lifecycle of a wind farm, the manufacturing and installation of wind turbines requires materials made from fossil fuels, requires water and energy, and consequently causes carbon dioxide emissions. At the end of the life of a wind turbine, most components or materials used in the construction such as metals and wires can be recycled and reused. The problem is the disposal of the fiberglass blades. Due to the length of these blades, they are very difficult and expensive to transport, and there are no commercially available technologies to separate the resins from the fibers or to make use of the material. This adds another environmental concern to the life cycle of a wind turbine. As we all know, wind comes and goes. Some days, it blows strong across the country, and the next day, there is no wind at all. This leads to the fact the wind farms should only be used in two ways. The first application is to use a large wind farm as complementary power generation facility to feed into a power grid that is primarily supported by a more available, steady energy supply such as natural gas. In that case, the natural gas demand will be lower during windy days when the wind farm feeds into the grid and carbon dioxide emissions will be lower during these days. The second application is the use of single or a low number of wind turbines for private, residential use. In this application, the power generated by the wind turbine(s) will be used to supply power to the residential area connected to the turbine. Access power can be sold into the public power grid or stored for future use. On days with low or no wind, the flow of power will be reversed, and the residential area will get power from the grid. It is very difficult to evaluate the economic part of a project like this and to justify the required capital investment. In some countries, the government pays subsidies to make these projects feasible. The question remains if it makes sense in the long run, and if achievable benefits justify the investment risk and negative impact. And a last word from an engineer's view, the efficiency of harvesting wind is low compared to other technologies. The amount of energy that can be transferred from the wind onto the turbine and generator is only a small percentage of the energy potential that the wind carries. However, in some cases it might still make sense, and we see wind energy as one of the solutions for small-scale, decentralized power generation.

Solar energy or solar power is the capture of energy from solar radiation and its conversion into thermal or electrical energy. The Sun is the star of our solar system. As far as we know today, it consists of about 70% hydrogen, 28% helium, 1.5% combined from carbon, nitrogen and oxygen, and 0.5% of neon, iron, silicon magnesium, and sulfur. The composition of the Sun changes very slowly over time as in its core, it converts hydrogen to helium and energy. This nuclear fusion reaction creates unbelievable amounts of energy in form of gamma rays. The radiation emitted by the surface of the plasma ball named Sun – we call it sunlight – is the force that allowed the development of life on our planet and that gives the Earth and everything on it the energy to survive. A solar power plant is a facility that converts the Sun's radiation into electricity. This conversion can happen either directly via photovoltaics or indirectly via thermal energy. Photovoltaic facilities use large areas of solar panels (photovoltaic cells) to capture solar radiation and converting it into electricity. The problem with these facilities is the footprint as larger power generation will require larger surface areas for panels. The cells or panels generate direct current (DC) power, so in almost all cases, multiple cells are combined to modules, and multiple modules are combined to arrays. Arrays are connected to inverters which produce the electricity and the desired voltage and convert to alternate current (AC) power at the desired frequency. The two main applications for solar panels are large solar power plants or small-scale applications, for example, on rooftops. In both cases, consideration must be given to the fluctuation in radiation over seasons and due to weather changes. Most systems include power storage in form of batteries and are connected to the public power grid as back-up. Decentralized units may also use a fuel-driven generator as back-up to the solar panels. The manufacturing of solar panels using silicon and thermal oxidation for surface passivation as well as the disposal of the panels at the end of their life cycle also needs to be considered when assessing the carbon dioxide emissions and environmental impact of this technology. The other type of solar power plant, the concentrated solar power plant, uses a system of mirrors and lenses to concentrate or bundle the sunlight and use its thermal energy to generate steam which then can be used to generate electricity with a conventional steam turbine and generator combination. For the transfer of the thermal energy and for its storage, these facilities use a working fluid. Like wind energy, solar energy will play a significant role in the development of low carbon intensity energy. We expect solar and wind or a combination of the two or of one of them with another energy source to be applied in the development of sustainable decentralized power supply solutions. They will also play a huge role in the development of smart cities that will allow to decrease the carbon footprint of humans significantly. Due to the known limitations in efficiency, energy storage and space requirements, neither wind nor solar power, will be the technology replacing fossil energy.

3.1.4 Nuclear

Probably, the most controversial non-carbon energy source is nuclear power. Nuclear power is a clean and very efficient technology that uses low-enriched uranium as fuel. The uranium atoms are split on a process called fission, and the large amount of thermal energy released during the process is used to generate steam which then is used to drive a steam turbine-generator plant to produce electricity. Fission was discovered in 1932 using lithium atoms. It was developed further, and the first man-made nuclear reactor reached criticality in December 1942. It was part of the Manhattan Project. The development of the light water reactor (LWR) and other reactor types allowed the commercialization of the technology and nuclear power plants reached capacities of up to 1500 MW per unit. Nuclear power plants have very low carbon emissions that rival the values of hydroelectric power. And nuclear power has the highest energy density of all power-generating technologies. In terms of available space – either as area required for a power plant or as volume of space occupied by the power plant – nuclear power provides by far the highest energy output compared to all other technologies. This makes nuclear power economically very attractive. There are discussions around the economics of nuclear power plant as the return of investment must consider the cost of construction as well as the cost of the decommissioning of the facility and the transport and storage of the radioactive waste. We still think that overall nuclear power has immense benefits over other energy sources. Unfortunately, these benefits are overshadowed by the negative headlines made by nuclear power. It started with concerns being raised very early in the development of nuclear power that were centered around the disposal of the nuclear waste. Concepts developed to solve the problem mostly involved burying the waste in caves or cavities deep enough to restrict the remaining radiation from harming humans or the immediate environment. And even more damaging to the image of nuclear power were the horrible accidents that happened in multiple facilities. The most commonly known accidents are the Three Mile Island accident from 1979, the Chernobyl disaster from 1986, and more recently, the Fukushima Daiichi nuclear disaster in 2011. The failure of safety controls in almost all cases supported by human error and complacency poses a risk that many countries are not willing to take. It must be seen if the pressure of decarbonization may allow nuclear power to experience a revival. The first wave of new nuclear power plants was expected in 2000– 2005, but it was stopped by the fact that there was very little knowledge in the construction industry around the high standards required to increase safety in nuclear power plants. One development that could allow nuclear power to make a comeback is the introduction of small modular nuclear reactors (SMNRs). The basic concept allows for the fabrication and assembly of advanced nuclear reactors with a capacity of up to 300 MW by using smart designs that simplify the plant assembly and allow for application of modular fabrication and construction methodologies. We predict that nuclear energy will at a minimum maintain its global market share with a good chance of even growing a few percent.

3.1.5 Crude Oil–Petroleum

Crude oil or petroleum is the fossil energy source with the second highest carbon intensity (coal having the highest carbon intensity). It is a naturally occurring liquid that was formed and is stored in certain geological formations beneath the surface of our planet. It has a yellowish dark brown to black color and can be found in various qualities, ranging from light to heavy oil (measured by its density), or from sweet to sour crude oil (measured by its sulfur content). Its composition is dominated by carbon (~85 wt%) and hydrogen (~12 wt%), with varying quantities of sulfur (0.05–6 wt%), nitrogen (0.1–2 wt%), oxygen (0.05–1.5 wt%), and various metals (<0.1 wt%). In its extreme form, crude oil can occur as tar or bitumen-like liquid mixed with sand and water. On the other side of the spectrum, the latest discoveries of shale oil are more considered condensate or very light crude oils that lack the heavy end of the full crude oil that is normally processed in a refinery.

Due to its transportability, high energy density, and relative abundance, crude oil has become the most significant and dominating source of energy and fuels on a global scale. In the past, up to 85% of the hydrocarbons contained in petroleum (by volume) were converted to fuels such as gasoline, diesel, jet fuel (kerosene), heating oils, other fuel oils, and liquified petroleum gas (LPG). Conversion process within refineries can be separated into two groups: hydrotreating, hydrocracking and fluid catalytic cracking are conversion processes that add hydrogen to the products; visbreaking, thermal cracking, delayed coking, and fluid coking are carbon rejection-based conversion processes. Due to the decarbonization of the energy and fuel market, a lot of people think that oil refineries will disappear as they are blamed to produce the majority of fossil fuels that we know of. However, all predictions and outlooks agree that this will not be the case. The hydrocarbon molecules coming from an oil refinery are used to make so much more than just fuels. In some integrated facilities, more than 50% of the products coming from oil refining go into the petrochemicals and chemicals sector as feedstock for their manufacturing processes. A lot of items we all use daily are derived from petroleum. Here are just a few examples:

- Medicine – over-the-counter medication, homeopathic products, vitamins
- Cosmetics – make-up, shampoo, perfumes, waxes, colors
- Plastics – all petrochemicals, plastic wraps, components for cell phones and other electronics, water bottles, and containers; this group alone takes 4%–5% of the petroleum molecules
- Synthetic rubber – shoes, tires, wet suits, gloves, and other PPE
- Cleaning products – detergents, solvents, and other ingredients
- Asphalt – also called bitumen; used as binder for road pavement (over 11 million miles of paved roads globally) and roofing material.

This list can be continued as it contains hundreds of other products that are made from petroleum and that we are sure nobody wants to miss anymore. We will provide a more detailed view on the refining markets in Section 3.2.

Crude oil prices have fallen since the beginning of 2020 due to reduced energy and fuel demand caused by the COVID-19 mitigation efforts. The Organization of Petroleum Exporting Countries (OPEC) and its partner countries agreed on reduced production levels for an extended period to stop the free fall of oil prices. And although oil prices seem to have stabilized for now, the current level is the lowest it has been in over 20 years. This causes uncertainties across all energy sources, including coal, natural gas, and renewables.

3.1.6 Natural Gas

Natural gas is the lightest of the fossil fuels and has the lowest carbon intensity. It is a hydrocarbon gas mixture containing methane (up to 95%) and varying amounts of ethane, propane, butane, carbon dioxide, nitrogen, hydrogen sulfide, and gases such as helium or argon. It can be found as a conventional resource stored in certain porous rock formations, as associated gas in pockets on top of crude oil reservoirs, or as unconventional gas or shale gas stored in shale rock formations that require a more advanced production technology called hydraulic fracturing or short fracking.

Of all fossil fuels, natural gas burns the cleanest with the least amount of carbon dioxide or GHG emissions. Therefore, natural gas is the choice of transitional energy source to reduce carbon dioxide emissions. During the next 20–30 years, natural gas will continue to replace coal and petroleum in the energy and fuel sector, and help achieving the first transition phase of decarbonization.

One of the challenges of dealing with natural gas is the volume of the gas that needs to be handled in storage and transportation. All currently applied solutions add cost to the use of natural gas and became only economically feasible with the drastic reduction in natural gas price based on the rapid growth of gas supply from shale plays. These are the most common methods to store and transport natural gas:

- Transport via gas pipelines and compressor stations; storage in pressurized bullets or tanks, which is constrained due to the high cost of high-pressure containers; however, compressed natural gas or CNG is a valid option for small, regional markets such as short-range transportation or residential heating.
- Liquefaction of natural gas to create liquified natural gas (LNG); storage and transportation in cryogenic vessels.

Table 3.2 will give you an idea of the abundance of natural gas based on shale plays alone.

Add In addition to the unconventional gas reserves energy companies have reported proven reserves of conventional natural gas of over 175 trillion m^3. This number increases as new fields are explored and new reserves are still being discovered, which is the reason for the confidence all energy companies have in the future of natural gas as transitional energy source. One of the challenges that remains is the commercial production of natural gas from the vast amount of small gas wells and reservoirs. In most predictions, only the large gas wells have been accounted for as it requires a large and steady flow of raw natural gas to justify the capital investment that is required for its production and distribution. Smaller wells and reservoirs are often abandoned, or worse, in some countries, the excess natural gas associated with the production of petroleum is flared at the well sites. This has become a big problem as these flares not only destroy valuable energy, but the combustion of the gas also creates significant amounts of carbon dioxide and other GHGs that are emitted into the atmosphere. Several initiatives are on the way to (1) make the production and

TABLE 3.2
Global Shale Gas Reserves – Confirmed

Ranking	Country	Shale Gas Reserves Confirmed (trillion m^3)
1	China	32
2	Argentina	23
3	Algeria	20
4	USA	19
5	Canada	16
6	Mexico	15
7	Australia	12
8	South Africa	11
9	Russia	8
10	Brazil	7
	Rest of the world	43
	Total shale gas	206

utilization of smaller gas wells and reservoirs economically attractive and feasible, and (2) to eliminate flaring of gas by applying proven, small-scale gas recovery and monetization technologies. These small-scale gas monetization technologies that are available or under development include simple gas compression and transmission to a centralized gas processing facility, onsite gas liquefaction (small-scale LNG), or onsite conversion of gas to a liquid form of fuel such as methanol or even diesel or a synthetic crude oil (via the Fischer–Tropsch conversion). The feasibility of small-scale gas-to-liquids (GTL) technologies strongly depends on market conditions and continues to fail. However, the increasing pressure from more stringent environmental regulations, especially around the sulfur content or content of other contaminants in fuels, may push these technologies over the edge into commercial applications. Another factor that will support this development is the ability of licensors and technology providers to decrease the capital cost of their equipment as well as the operating expenses for the completed facility. This way small-scale GTL will become more competitive to other technologies. This will allow investors to choose the right technology and product for the specific local market and economic environment they are working in.

We already established the connection between oil refining and petrochemicals and the use of certain petroleum-derived products as feedstock for the petrochemical industry. To complete the picture, we must acknowledge that natural gas, due to the commercialization of hydraulic fracturing and the resulting abundance of natural gas, has become a competitor to petroleum as a feed source for the petrochemical industry. Natural gas and the associated natural gas liquids (NGLs) contain ethane, which can be used to replace naphtha as feed to steam crackers for production of ethylene and propylene. While some cracking furnace designs allow for the use of different feedstocks that may range from gasoil via naphtha to ethane. In most cases, the furnace must be modified to allow for a change in feedstock quality. And the operating parameters must be adjusted to account for the difference in cracking reactions that will occur inside the furnace tubes. Other sections of the steam cracker unit that may need modification due to change in feedstock are the quench oil column or the demethanizer column. In the past years, due to the rise of shale gas many operators along the US Gulf Coast either decided to convert their steam cracking furnaces from naphtha to ethane or, for newly constructed capacity (expansions or green filed), decided to design for ethane as the preferred feedstock. Even operations in Europe conducted studies to evaluate the conversion of their furnaces to ethane cracking. The economics of feedstock selection can not only be driven by its price, but also by the yields that can be achieved. As can be expected, feeding ethane into your steam cracker will give you by far the highest yield in ethylene. However, this comes at the cost of propylene yield and output of other by-products such as hydrogen and C_4-hydrocarbons. If your downstream value chain requires any of the products besides ethylene, naphtha is the better feedstock. As long as natural gas prices, and associated with it the price of ethane, stay low and competitive to naphtha, ethane will maintain its position as viable feedstock to the petrochemical industry. However, the recent events around COVID-19 in combination with the oil price war have benefited naphtha up to a point where the margin from using naphtha as a feedstock in hydrocrackers has surpassed ethane for the first time in 2–3 years.

3.2 MARKET OUTLOOK REFINING

A typical petroleum or crude oil refinery will sell the following products to local or foreign markets:

- Gaseous Products
  - Methane – often used internally as fuel gas
  - Ethane – often used internally as fuel gas
  - Propane
  - Butane.
- Liquid Products
  - LPGs – propane and butane
  - Chemical naphtha
  - Gasoline – regular and premium
  - Kerosene – mostly as jet fuel of different grades
  - Diesel – ultra-low sulfur quality (ULSD) and regular
  - Low sulfur fuel oil (LSFO)
  - Bunker fuel – see comments on IMO 2020.
- Specialties and Solid Products
  - Aromatics – benzene, toluene, xylene (BTX)
  - Asphalt
  - Petroleum coke – fuel grade, anode grade, specialty grade (needle)
  - Sulfur – liquid or solid.

For most refineries, their margins and economics are driven by the liquid fuels' markets. Gasoline, diesel, jet fuel, and bunker fuel prices in relation to crude oil prices determine the economic performance of oil refineries. There are always exceptions to the rule. Some refineries produce other products for a well-defined market, for example, needle coke for the electrode and graphite industry, that may have a big impact on the economics of this specific refinery. However, these cases are rare, and all other refineries must face the challenge of competing in the liquid fuels' markets.

The outlook for the global energy demand has changed over the past 5 years as projections are not as optimistic as they had been before. We agree with the more realistic view and predict that growth in global energy demand will decelerate to 0.5%–1.0% per year through 2050, which is a rate that is about 30% slower than previous forecasts. This view is supported by the following developments.

- Emerging and developing countries such as China, India, Latin America, and African countries will drive all growth in energy demand, while in European and North American countries, the growth in energy demand will decline. In these countries, the energy consumption will grow slower than the GDP due to efficiency improvements and implementation of work processes with lower energy intensity.
- The growth of fuels demand will slow down. In countries such as the USA where the production rate of crude oil and natural gas will increase at the same time, the export of crude oil, refined products, and LNG will increase.

- Demand for chemicals will grow at more than double the rate of total energy demand.
- Demand from individual and regional transportation will peak around 2023.
- Demand for electricity will grow much faster than the demand for other energy sources by more than two to one.
- Solar and wind energy will represent almost 80% of all added energy generation capacity and may grow its share to 34% of global energy generation by 2050, supported by federal tax credits and regional programs to increase the decentralized use of these technologies.
- Fossil fuels will keep dominating the total energy blend through 2050. However, their share of total energy will decline to about 75%.
- As the transitional energy source of choice, natural gas demand will grow at almost double the rate of total energy demand.
- Due to the use of coal to cover the immediate energy demand in developing countries, it may peak by 2025, but will experience a steep decline in all European countries and in North America.
- The use of biofuels in the gasoline, diesel, and jet fuel blends will increase. For example, the content of biofuels in transportation fuels is expected to rise from currently about 7% to 9% within the next 12 months. In more optimistic predictions, this value can reach 13.5%.
- Oil demand growth will flatten to a value slightly above zero and flatten out on a global scale. In some areas such as Europe, the demand will show a slight decline.
- Carbon dioxide emissions from energy generation will flatten and start to decline around 2035. This is mainly driven by implementation of technology improvements in cars and light vehicles, for example, the development of more efficient combustion engines and the increased sale of electric vehicles. Another factor will be the shift to wind and solar in power generation.

Based on the developments listed above, the total demand for liquid hydrocarbons from refineries will be determined by the balance between the growth predicted for the petrochemical sector and the declining demand in the private transportation sector. The demand for petrochemical feedstock will drive up to 70% of the growth in demand for liquid hydrocarbons through 2035. The demand for liquid fuels such as gasoline and diesel will peak in the next few years and flatten by 2025 as the private transportation sector will benefit from the higher efficiency in combustion engines and the added numbers of electric vehicles. This will mainly be implemented in fully developed countries. Emerging and developing countries will still have a higher market share of less efficient and non-electric vehicles as the private transportation sector in these countries will be lagging other countries by 10 to 15 years or even more. Figure 3.1 shows the past development of the global liquid fuel demand.

In 2020 the price for Brent, one of the crude oils used at benchmark value for the petroleum markets, will average at about US$ 34 per barrel. In 2021, this value will go up to about US$ 48 per barrel. All major investment decisions for maintenance, revamps, optimization, and new capacities will be based on this scenario. Long-term predictions see the price for Brent at about US$ 105 per barrel by 2050.

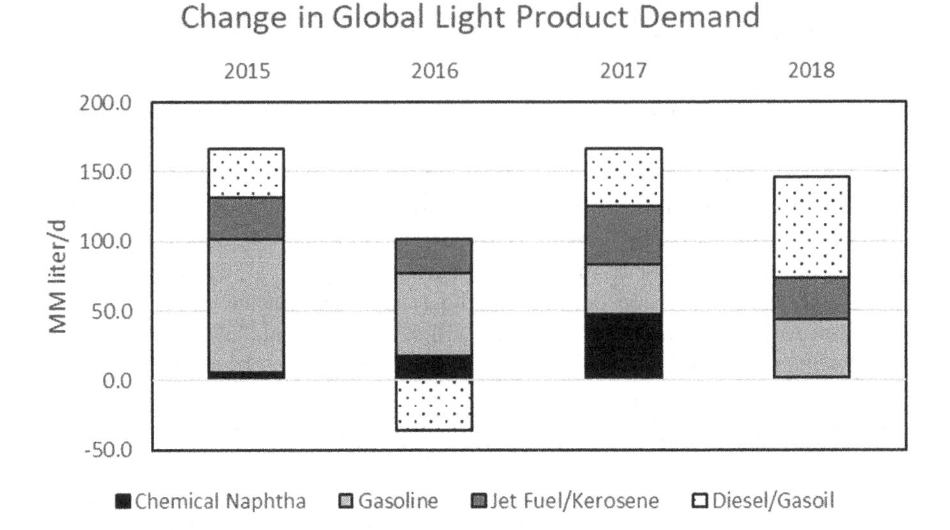

FIGURE 3.1 Past development of global liquid fuel demand.

Due to the uncertainties and the challenges in predicting all influencing factors, the range attached to this value is from US$ 46 to US$ 183 per barrel.

Environmental regulations are expected to become more stringent in the future. Government agencies will keep moving the limits for sulfur in liquid fuels as well as the limits for other components such as benzene and aromatics. This will put more and more financial pressure on refiners since achieving these limitations will become more and more challenging technically. The cost of producing these low impurity fuels will drive all refiners toward finding new outlets for their products. The logical option is the integration with a petrochemical complex. Exchanging products with a petrochemical complex will provide a huge cost benefit to refiners in the future. This trend will also help with the response to changing regulations on aromatics and benzene content in gasoline. The extraction of benzene, toluene, and xylene from reformate makes more sense in the context of petrochemicals than fuel production.

Refineries in Europe, the Middle East, and North America will be under pressure of declining margins as they serve fully developed markets that, in long term, will experience a slow decline in the demand for liquid fuels. Achieving a higher degree of integration with the petrochemical sector will be the main strategy as it will also help the petrochemical manufacturers with meeting the increased demand for stable production of olefins and other products feeding the plastics sector and chemicals sector. In countries with high level of integration between refining and petrochemicals, the utilization of refinery assets will be stable at levels between 90% and 93%. The highest margins will be achieved by refineries with the capabilities to produce and supply ultra-low-sulfur diesel (ULSD) and low-sulfur bunker fuel and fuel oil (LSFO). Figure 3.2 shows the regional refining utilization forecast.

Please note that in the chart, Europe represents the UK, the Netherlands, and Belgium, North America represents the US Gulf Coast, and Asia represents Taiwan,

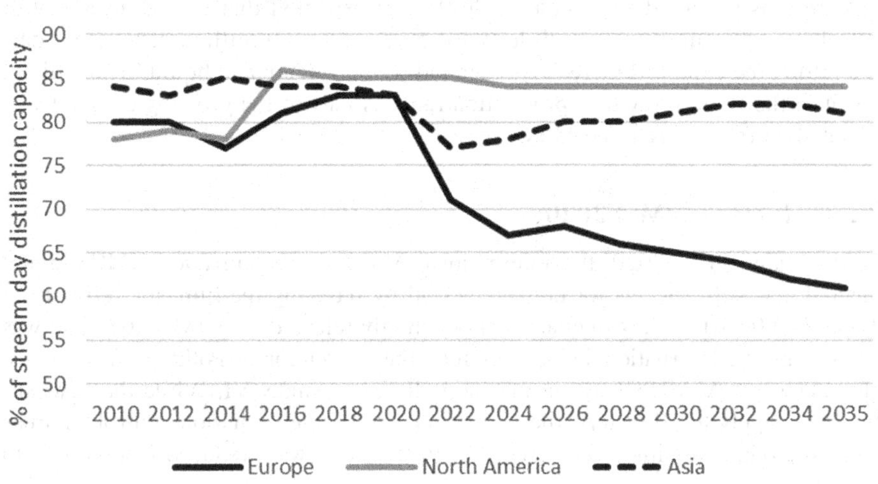

FIGURE 3.2 Regional refining utilization forecast.

Singapore, and South Korea. In 2020, the consumption of petroleum and liquid fuels will decrease by almost 6% but recover in 2021 by showing growth of about 7.5%.

The shale plays in the USA allowed the country to change from a net oil importer to a net oil exporter. The growth in production of tight and unconventional oils exceeded all expectations and was only stopped by COVID-19 and the resulting decline in fuel demand and free fall of oil prices. Only once this situation normalizes, it will be clear if the trend will continue or if permanent change has occurred. So far, the only limitation to the growth of the US oil market was the infrastructure to store and transport the oil. Investments in new tank farms and pipelines couldn't be made fast enough to keep up with the growth in oil production. And some projects couldn't pass the hurdle of environmental approvals. The US Gulf Coast refiners took advantage of the light oil by investing in new process units to replace heavy sour crude with light and mostly sweet condensate. However, having heavy oil process units sitting idle or operating at low capacity is economically not feasible. Due to the political situation in Venezuela and its failure to maintain its oil production and refining assets, the country with the highest oil reserves in the world now must import refined products to support the domestic demand for fuels. This opened the doors for heavy Canadian crudes to enter the US Gulf Coast markets and supply the much-needed heavy oil balance to the US refiners. The forecast still stands that most US oil exports – if not all – will go to Asia to support the growing markets there.

The retail prices for all grades of gasoline will show only moderate growth on a global scale. Regional spikes may occur based on supply and demand interaction. Retail prices for diesel, especially ULSD, will show a slightly higher growth compared to gasoline since diesel is considered part or the fuel mix for long haul and heavy transportation needs.

In Q1 of 2020, the capacities for storing petroleum and petroleum-derived products reached limitations on a global scale as inventory of these liquids by over 2.5 million barrels per day due to lower demand caused by COVID-19. As this trend is expected to reverse in 2021, it is expected that global inventory of petroleum and liquid fuels will decrease by close to 2.0 million barrels per day. Therefore, the expected increase in demand for petroleum and liquid fuels will not result in the same growth of production rates as part of the growing demand will be satisfied from stored inventory.

3.2.1 Impact of IMO 2020

Starting January 1, 2020, the International Maritime Organization (IMO) implemented new sulfur levels for bunker fuel oil by reducing the limit for sulfur from 3.5 wt% to 0.5 wt%. This regulation is commonly referred to as IMO 2020 and was adapted by the International Convention on the Prevention of Pollution from Ships, also known as MARPOL (from marine pollution), Annex VI. While the new sulfur levels apply to all waters, the regulation also defines emission-controlled areas (ECAs) along coast lines for which the sulfur limits were reduced from 1 wt% to 0.1 wt% in 2015.

The marine sector accounts for about 50% of the global fuel demand, so the impact of the new regulations is significant. One of the obvious consequences is the increased cost for the low-sulfur bunker fuel which will result in increased freight cost for transportation of goods around the globe. IMO 2020 is also expected to have a long-term effect on the light/heavy differential and, of course, global refining margins.

The refining and marine industries had and still have the following options to react to IMO 2020 bunker fuel sulfur regulations:

- Installation of exhaust gas cleaning systems (EGCSs, also called scrubbers) to remove sulfur oxides from the exhaust flue gas on board the marine vessels
- Switching from high-sulfur fuel oil to low-sulfur fuel oil; this can be low-sulfur gasoil, low-sulfur residuals, or hybrids
- Switch to alternative fuel sources, most likely LNG, which requires CAPEX intensive modifications to the marine vessel and fuel supply infrastructure
- Non-compliance, either sanctioned due to acknowledgment of supply-side difficulties (transitional, category-based exemptions, local non-availability, etc.), or gray area non-compliance (i.e., malicious intent).

For owners and operators of marine vessels operating outside of the ECAs, there was very little benefit in pre-investing in the installation of scrubbers. And based on the ECA change that occurred in 2015, there was a strong indication that marine vessel owners and operators will likely wait until after the implementation of IMO 2020 before reacting. Early estimates showed a slow initial ramp of scrubber installations before 2020 of about 100 scrubbers per year, which was probably close to the actual number. The forecast for the price spreads between low- and high-sulfur bunker fuels supported the emergence of scrubber financing schemes that would support an increased number of scrubber installations. However, this forecast needs to be revisited after the analysis of the long-term impact of COVID-19. By next year, experts estimate that there will be about 2000 less scrubbers installed than previously forecasted due to reluctant adoptions, limits on scrubber installation capacity, and reduced payback on scrubbers.

The forecast for refining margins included a substantial positive change from 2019 to 2020 as a result of the IMO 2020 implementation. Again, this was disrupted by COVID-19 and the long-term effect still needs to be analyzed and verified. In general, long-term refining margins will have a long, steady decline after IMO transition period defined as the years 2020–2024 due to decline of the global products demand growth based on higher efficiency engines and use of alternative or renewable fuels.

3.2.2 Changes in Private Transportation

In the next 10 years, electric vehicles including hybrid vehicles could represent up to 30% of new cars sold globally, with a market share up to 50% in China, the European Union, and the USA. The next forecast will also have to include autonomous-vehicle adoption and car sharing. If the acceptance of electric, autonomous, and shared vehicles continues to grow as it has in the past 2 years, oil demand based on fuel demand from the private transportation sector could be about 3 million barrels lower in 2035 than assumed in the current forecast. Accelerated adoption of light-vehicle technologies and the adjustment of plastics demand could reduce the global oil demand by nearly 6 million barrels per day by 2035. Global oil demand is expected to peak around 2030, but at a lower value than predicted in the past years. In some scenarios, it will not reach the benchmark of 100 million barrels per day.

Innovation in the design and manufacturing of large format batteries will determine the rate at which electric vehicles will grow in market share as the cost for the battery pack is one of the big items in setting the price for the vehicle itself. In 2007, lithium-ion storage batteries cost about US$ 900 per kilowatt-hour. In 2015, the cost was already lowered to about US$ 380, and it is on track to drop below US$ 200 this year.

But even with the increasing forecast for the share of electric vehicles among new cars being sold, they have barely passed the 1% mark globally. This is mostly contributed to the price of the vehicles, the lack of confidence in the charging station infrastructure, and the failure of government subsidies to attract people to buy electric vehicles. The use of autonomous trucks in mining and farming may become a factor in delivering big savings on labor and carbon dioxide emissions. The adaption of car sharing services is slowly taking off in Europe and the USA, while Lyft, Uber, and others have upended the taxi business and begun to change patterns of personal vehicle ownership and public transportation choices.

3.2.3 Long-Term Outlook

The human population will continue to grow globally and will cause an increasing demand for energy and fuels. The first reaction could be that we need more refineries to be built globally to meet the increasing energy and fuels demand. That was the justification for the construction of new refining capacity in the Middle East and Asia. Other regions such as Europe have excess capacity, but the cost of transportation and the low margins make the supply of fuels to emerging markets prohibitive.

In parallel to the growth in overall population, the global population is aging. By 2050, about 25% of the population of developed economies, including China, will be 65 or older. The proportion of workers in the total population will decrease, and some experts predict that the shrinking labor force will lead to a global macroeconomic downshift. If the current trends continue and do not account for any unexpected increase in productivity, the growth in Gross Domestic Product (GDP) could be 40% lower in the years after 2050 compared with the first half of this century.

Additionally, it can be expected that the structure of GDP growth is shifting toward services. Following in the footsteps of fully developed economies such as Europe or North America, developing and growing economies such as China will shift their focus away from heavy industry to services to keep their economies growing. At the same time, the surge of energy-intensive industrialization that we have seen in growing economies during the past decades will likely not be replicated in emerging economies. Their early shift to service industries means a greater share of global GDP will be driven by less energy-intensive industries. Accounting for all sectors of the global economy, the energy intensity of global growth will decline by about 50% by 2050.

There are other factors that will influence the refining margins in the next 20–30 years. As we have seen in the past, oil prices could decline and lead to an increased demand of fuels. During COVID-19, we have no learned that there is a scenario that includes decreased demand and very low oil prices. The acceleration of technology development and adoption will play a huge role in the development of alternative and renewable fuels. Individuals as well as small, even large businesses could change their behaviors and implement workflows and technologies that are more energy efficient. This could be supported by changes in policies and regulations that could realign financial or fiscal incentives for suppliers and consumers (for example, revised carbon taxation).

Photovoltaic installations and installations of wind farms have accelerated the growth of the energy market share for solar and wind technologies. This trend will continue over the next 20–30 years as even considering the uncertainties around the progress of technology development, the cost for these installations will continue to decrease.

Unconventional oil and gas will also continue to grow and establish a higher energy market share as new reservoirs will be developed and production will grow on a global scale.

Previous forecasts had put a penalty for high adoption cost and slow development on innovative technologies. This was proven to be a wrong approach as development happened a lot faster than expected and the cost of more efficient technologies – for example, the reduction in cost for LED light bulbs – was reduced

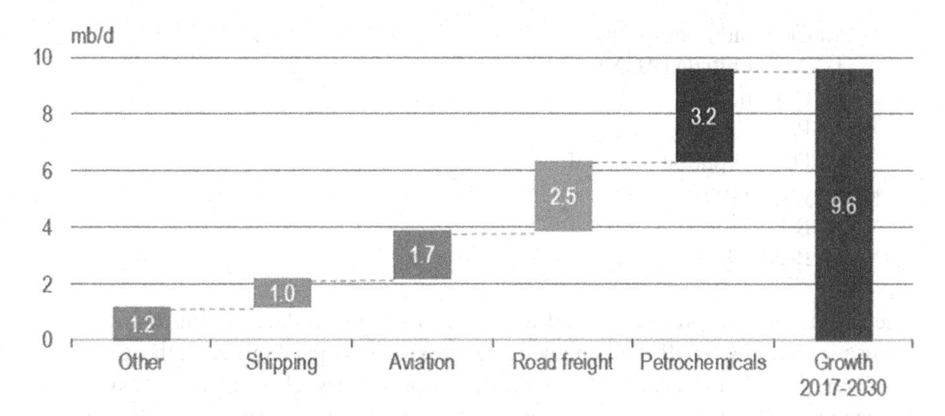

FIGURE 3.3 Refining growth source distribution.

enough to change consumer behavior and attract more people to the new, more energy-efficient technologies.

The estimated sources for growth in the refining sector are shown in Figure 3.3.

3.3 MARKET OUTLOOK PETROCHEMICALS

The products coming from the petrochemical industry are very diversified and can be grouped as follows:

- Olefins
  - Ethylene
  - Propylene
  - Butadiene.
- Polymers
  - Polyethylene (PE) – high density (HDPE), low density (LDPE), ultra-low density (LLDPE)
  - Polypropylene (PP)
  - Polyvinyl chloride (PVC).
- Aromatics
  - Mixed xylenes (MX)
  - Paraxylene (PX)
  - Orthoxylene (OX)
  - Benzene
  - Styrene
  - Toluene
  - Methyl tert-butyl ether (MTBE)
  - Purified terephthalic acid (PTA)
  - Polyethylene terephthalate (PET)
  - Monoethylene glycol (MEG).
- Methanol and other alcohols
- Ammonia and fertilizers

- Solvents and intermediates
  - Acrylonitrile (ACN)
  - Oxo-alcohol
  - Phthalic anhydride (PA)
  - Dioctyl phthalate (DOP).
- Recycled plastic
  - R-PET
  - R-HDPE.

The petrochemical products listed above are used to produce a long list of items we use daily, such as all items made of plastics such as packaging, clothing or toys, fertilizer, digital devices, detergents for dishes or laundry, cleaning agents, synthetic rubber products such as tires, and even medical devices. Petrochemicals are also found in many parts that are needed to build and operate modern energy generation systems, for example, solar panels, wind turbine blades, batteries, thermal insulation, or parts for electric vehicles.

Especially, the demand for plastics has grown significantly driven by the growth rates in fully developed economies. Similar numbers can be seen for other products such as fertilizer. If similar growth rates can be expected from the developing and emerging economies, there is no surprise that the long-term outlook for petrochemicals is very positive and shows a more steady than cyclic nature. Plastics also outpace other bulk materials required to build economies such as steel, aluminum, and cement.

In the recent past, the USA and China have seen the largest petrochemical capacity additions. The long-term forecast shows the next capacity additions to happen in other Asian countries and the Middle East. The Middle East and the USA have a feedstock advantage as they have access to low-cost ethane from their abundant natural gas supplies. This advantage allows both regions to gain the lion's share of ethane-based chemical exports in the short and medium terms. The added capacities put pressure on the prices in the respective markets, which is multiplied by the trade tensions between the USA and China. In the short term, this may lead to rerouting of product streams and a change in the strategy for future investment in additional capacities. Lower prices in the petrochemical sector will also lead to incentives for pushing aromatics into the gasoline blending pool.

Coal-based methanol-to-olefins capacity in China is expected to grow 100% between now and 2025. This is a major building block for providing the material inputs for its large domestic manufacturing industry. In the longer run, other Asian countries and the Middle East both are forecasted to increase the market share of high-value chemical production by 10%. The share coming from Europe and the USA will decrease in the same period. By 2050, India, Southeast Asia, and the Middle East together will account for about 30% of the global ammonia production.

Increased recycling and stronger efforts to reduce single-use plastics will become more successful, especially in Europe, Japan, and Korea. These efforts will be outweighed clearly by the steep growth rates for plastic consumption as well as its disposal in developing and emerging economies. It will also be challenging to find alternatives and alternative sources for plastics, which supports the forecast that the global demand growth for petrochemical products will stay strong.

The only challenge for oil-based petrochemical production will come from shale gas plays and the abundance of natural gas in certain regions. After almost 20 years of stagnation and declining production rates, the USA has returned to growth as a low-cost manufacturer for chemical production due to the shale gas revolution. Currently, the USA accounts for about 40% of the global capacity to produce ethane-based petrochemicals. Thanks to Saudi Arabia and Iran, the Middle East remains the low-cost leader for key petrochemicals. Consequently, a significant number of new projects have been announced across the region.

For naphtha-based petrochemical production and high-value chemicals, China and Europe each account for around a quarter of the global capacity. China's coal-based chemical industry developed from a once a speculative proposition through steady technological improvements to an important factor in the growth of the Chinese economy. India is expected to grow their petrochemical industry and leave its current level of only 4% of global capacity, mainly driven by growth in its domestic demand.

Historically, the petrochemical markets were strong, but always subject to cycling due to shift in supply and demand structure. This forced petrochemical companies to apply creative strategies to protect and stabilize their businesses. This can be achieved by securing a stable supply of feed streams and by optimizing the supply and value chain for each of the product lines. For many companies, increased productivity and functional excellence – in other words, executing a business model better than most competitors in the field – will become the key to running a successful and growing business. At the same time, the development of a strategy will also become much more difficult as it will be more challenging to identify the remaining opportunities for growth that exceeds GDP and to develop approaches that will allow the conversion of those opportunities in a value-generating way.

In the past, the petrochemical and chemical industries' traditional rule of thumb was to assume that the demand would grow at 1.3–1.4 times the rate of GDP. The rate will change globally as mature markets reach a saturation point for plastics. For example, markets such as Germany and Japan are declining in per capita plastics demand. The new forecast expects chemical demand to grow at only 1.2 times the GDP in the short term. In the long term, that growth will decline to match the GDP growth rate. Two elements could transform chemicals demand further: plastics recycling and plastic-packaging efficiency. To play out such a scenario, if the global plastic recycling rate improves from currently 8%–20% in 2035, and if at the same time, plastic packaging use declines by 5%, the estimated demand for liquid hydrocarbons as driven by chemicals could be about 2.5 million barrels per day below the current forecast. It is easy to understand that the application of circular-economy principles to the plastics markets could dramatically reshape the economics of this workhorse of the global economy.

The petrochemical industry is also challenged by climate change, air quality, and water pollution issues. Petrochemical products include a growing number of applications in various clean technologies critical to a sustainable energy system. However, the production, use, and disposal of these products pose several sustainability challenges that need to be addressed. The chemical sector consumes roughly the same amount of energy as the steel and cement industries combined. But it emits less carbon dioxide than either sector. Still, the emissions from the petrochemical industry amount to around 1.5 $GtCO_2$, which is 18% of all industrial-sector carbon dioxide emissions, or 5% of total combustion-related carbon dioxide emissions. The chemical industry consumes more oil and gas than other heavy industries, which tend to rely more on coal. And the carbon contained in chemical feedstocks is mostly locked into final products (such as plastics) and released only when the products are burned or decompose. Almost 35% of all plastic packaging leaks or is disposed of into our sewer systems, resulting in about 8 million metric tons annually polluting the oceans. To address this issue, the industry is looking at mitigation strategies such as developing incentives for recycling, reuse, and controlled disposal. However, with today's designs, technologies, and systems, these strategies can only be partially realized. A recent study found that 53% of plastic packaging in Europe could today be recycled "ecoefficiently." While the exact figure can be debated and depends on other factors such as oil prices and market developments, the actual message should not be a negative one, but should highlight the opportunities for improvements to be captured now to develop attractive targets for the global value chain to drive innovation toward a circular economy.

3.4 MARKET OUTLOOK NATURAL GAS

As we had mentioned earlier, natural gas is being brought to market either as CNG or as LNG. Heavier gases such as ethane, propane, and butane are marketed separately either in liquified form or as gases. Any heavier components are sold as NGLs or natural gas condensate. The dominant form of natural gas marketing is LNG.

As we have shown in section 3.1.6, the global proven reserves for conventional natural gas are around 175 trillion m^3. Together with about 206 trillion m^3 of unconventional (tight or shale) gas the energy companies report a total proven reserve of natural gas of over 380 trillion m^3 or 13,420 trillion ft^3. And to make the deal even sweeter, about 70% of these resources are estimated to be available for production at an average breakeven price of less than US$3 per million BTU. And last, but not least, use of natural gas is the main strategic component for the long-term reduction the carbon dioxide emissions and the transition to low and ultimately zero-carbon energy and fuels. These are the reasons why a lot of capital will keep flowing into the production, liquefaction, transport, and use of natural gas over the next decades.

In 2018, about US$ 360 billion of investment went into building natural gas infrastructure globally. Although this is a huge number, experts estimate that investments will need to be about US$ 450 billion or higher to meet the growth in demand for natural gas across all regions and industries. The fastest growing demand is coming from Asia, mainly supported by the limited access to their own natural gas resources. The main shortcomings in investment are in the midstream and downstream infrastructure. Old pipelines and aging gas processing plants will require substantial investment to deliver the required capacities and reliability to meet the future market needs.

The two main suppliers for natural gas will remain Russia and the USA, who had about 80% of the growth in natural gas supply in 2018 and will maintain that rate in the short- and long-term forecast. The growth in gas consumption comes from China and the USA, who shared about 60% of the global growth in 2018 and are poised to be the drivers for growth in the natural gas sector over the next 5–10 years. Other countries will have to focus their efforts more on natural gas to keep up with the pacesetting economies. This will require innovation to reduce cost of production and processing, infrastructure, and supporting elements like policies. Otherwise, the growing and emerging economies will keep using other cheap and readily available energy sources such as coal and oil to grow their economies.

The current forecast models show that up to 2050, the price for natural gas will be relatively stable at competitive levels. Using the US Henry Hub price as a reference, natural gas will be traded between US$ 3 and 4 per million BTU in 2050. That is a moderate increase from current prices of about US$ 2 per million BTU, considering the growth in demand over the next 20–30 years.

Another challenge is the establishment of a global LNG market and the balancing of LNG export and import capacities. LNG is slowly developing into a trade commodity that shows a certain price security and stability, which should encourage more countries to consider the import of LNG to supply their energy and power demands. Main additions to the global gas liquefaction capacity happened in Russia, Australia, and the USA. Global utilization of these capacities is still low at about 80%. To give the LNG market more stability and flexibility, more capacity will need to be added. In parallel, the LNG suppliers will need to adapt strategies to support the next generation of LNG users by considering more short-term contracts and smaller delivery volumes.

In summary, security of LNG supply at competitive prices is the key for a successful transition to a sustainable supply of lower carbon intensity fuel to established and growing economies. LNG will have to be able to join pipeline gas and other domestic gas supplies to compete with and replace coal and oil in the power and fuel sectors. The price of energy is often underestimated in the assessment of energy markets and selection of sources. It seems obvious that people living in growing or emerging economies will need to save money and can only afford the cheapest form of energy they can get access to. But recent events have shown that also in established and fully developed economies, there is only very little tolerance toward higher energy prices. While some governments thought that they could enforce the use of alternative and renewable energy by shifting the cost for development and infrastructure to the consumer, they were proven wrong and will need to rethink their strategy to successfully achieve their goals for reduction of carbon dioxide emissions. Stable and affordable LNG supply will be a key factor in that transition. It has been reported that in countries such as China and India, the cost for natural gas as fuel and energy source is becoming more competitive to coal. If one would add the cost of dealing with the higher carbon dioxide emissions from coal, natural gas would already have the advantage over coal in key growing and emerging markets.

In summary, the energy sector is determined to follow the path of decarbonization of fuels and energy to deal with the challenges the industry is facing. Decarbonization is the process of phasing out the use of fuels with a high carbon intensity such as coal and crude oil and switching energy and fuel manufacturing to low or zero-carbon sources. This process will take a very long time for obvious reasons. One of the reasons is the time it will take to find the "new" technology that will allow the use of zero-carbon, renewable, or other fuels economically in a large scale. Wind and solar are playing a role in this scenario and will still grow over the next decade. Battery-based electric vehicles will also be a factor, but the required generation of electricity makes this technology only partially independent from fossil fuels. And the high demand for lithium and graphite to manufacture high-efficiency batteries raises some concerns regarding the environmental impact these technologies have. There is a clear winner in the race for interim energy source: natural gas. While natural gas is still a fossil fuel, its carbon intensity is far lower than for coal and oil. It burns a lot cleaner and reduces emissions significantly. Transitioning from coal and oil to natural gas-based fuel and energy production will help with achieving interim goals for the desired reduction of emissions and slowing down the effect of global warming. For the next 20–30 years, natural gas will be the preferred source of fuels and energy. Wind, solar, and renewables will take a greater market share, but will not be able to replace fossil fuels.

3.5 MARKET CHALLENGES AND OPPORTUNITIES

The assessment of the energy markets and the outlook on forecasted developments in the different energy sectors leads to the realization that the refining industry and the petrochemical industry will face the following challenges in the next 20–30 years:

- Refining Industry
  - Manage and support the decarbonization of the global energy and fuel mix
  - Find alternative outlets for liquid fuels as markets will shift away from fossil fuels.
- Petrochemical Industry
  - Meet the global growth targets
  - Outperform the competition
    - Increase efficiency or processes
    - Achieve operational excellence.
  - De-cycle the global market and stabilize the supply and demand curves
    - Find secure and high-quality feed sources
    - Find outlets for by-products.

All petrochemical products are manufactured using these four feed streams: petroleum, natural gas, coal, and biomass. And about 50% of the energy consumption in the petrochemical sector comes from fuels derived from fossil energy sources. Up to 2030, the fuels demand from the petrochemical sector will account for 30% of the growth in petroleum demand, and by 2050, it will reach about 50%. By 2030, the petrochemical sector will also consume a considerable larger amount of natural gas than today.

Oil companies are increasingly pursuing integration with facilities along the petrochemical value chain. Against a backdrop of slower liquid fuels demand growth in contrast to robust growth prospects for petrochemical and chemical products with attractive margins, major oil companies are looking at further strengthening their links with petrochemical markets. New, direct crude oil-to-chemicals process routes may also come into play, offering alternatives to traditional refining/petrochemical operations although the technology remains challenging for now. For example, Saudi Aramco and SABIC have recently announced a large crude-to-chemicals project with a capacity five times the size of the only existing facility in Singapore.

Current refinery configurations apply technologies that are suitable to support the production of petrochemicals, for example,

- Production of a light naphtha stream labeled as chemical naphtha as feed stream for steam cracking units to produce olefins and aromatics
- Continuous catalytic reformers (CCRs) for aromatics and hydrogen production
- Fluid catalytic crackers (FCCs) to produce olefins and aromatics.

There are options to change existing refinery processes to maximize the conversion of crude oil fractions to petrochemicals. By changing the operating parameters and adjusting feedstock qualities, an FCC unit can be operated to yield over 50% of its products into petrochemicals. Currently, a typical value would be about 15% of products. The use of advanced reforming catalysts in a catalytic reformer will allow the refinery to increase its yield of benzene, toluene, and xylene (BTX) aromatics by as much as 7% over a typical catalyst currently used to maximize the yield of high-octane gasoline blend stock.

Especially for refineries operating in markets with suppressed or narrow refining margins, which soon may apply to almost all refineries, becoming part of an integrating processing hub for fuels and petrochemicals will bring a competitive edge by

- Spreading the business risk across multiple markets
- Reducing gas consumption and CO_2 emissions
- Improving flexibility
- Achieving higher efficiency
- Driving technology upgrades
- Creating jobs.

Petroleum refining and petrochemical industry coordination and integration gained a great deal of interest among investors and operating companies. Main drivers are the opportunity to enhance operational efficiencies, the higher return on existing and new assets through increased stream integration, and flexibility in operations and being able to meet the more stringent clean fuel regulations. Refineries will be able to take advantage of strong petrochemical markets. For petrochemical companies, the primary integration driver is competitiveness.

3.5.1 Trends of Integration by Region

Based on the level of integration between an oil refinery and a petrochemical complex, measured as a percentage of crude oil on a weight basis converted to petrochemicals, we propose the following classification:

- <5 wt% – low level of integration
- 5–10 wt% – moderate level of integration
- 10–25 wt% – high level of integration
- >25 wt% – petrochemical refinery (refinery dedicated to supplying products to the petrochemical complex).

The assessment of level of activity in the development of integration between refineries and the petrochemical industry considers the overall market share, the blend of feed streams used for petrochemicals, and the level of capital investment that is forecasted for each region.

3.5.1.1 North America

The North American region includes Canada, the USA, and Mexico. These three countries combined take about 14% of the global petrochemical feedstocks with the following distribution:

- 50% Ethane
- 15% Naphtha
- 30% Other oils
- 5% Natural gas.

The integration level between the North American refineries and the petrochemical industry can be estimated as moderate to high. After the massive wave of capacity additions along the US Gulf Coast and near the shale gas plays, most of the North American petrochemicals will be based on ethane as feedstock to the steam cracking units. Investment in integration with oil refineries will be low to moderate as it will focus on reliability improvements and feedstock flexibility.

3.5.1.2 South America

The South American region also includes Middle America and the Caribbean islands. This region accounts for only 4% of the global petrochemical feed market. The distribution of feed streams is as follows:

- 10% Ethane
- 50% Naphtha
- 5% Other oils
- 35% Natural gas.

The integration level in South America is high, but very limited investment is forecasted for integration activities in this region as the oil industry struggles and investors have their focus on the development of the South American shale gas plays.

3.5.1.3 Western Europe

The refining industry in Western Europe always had close ties to the petrochemical industry as facilities were located close to each other which allowed the exchange of intermediates and products to optimize performance of the facilities. The level of integration therefore is moderate to high. Western Europe consumes about 16% of the global petrochemical feed streams with the following distribution:

- 5% Ethane
- 65% Naphtha

- 20% Other oils
- 10% Natural gas.

The planned investment in this region for integration projects is low to moderate as most facilities already have a high degree of integration and additional investment will be difficult to justify. Any new investment will target reliability issues and feedstock flexibility.

3.5.1.4 Eastern Europe/Russia

This region only consumes about 4% of the petrochemical feed streams as the countries in this region, predominantly Russia, are focused on exporting petroleum and natural gas. The existing petrochemical plants are supported by

- 2% Ethane
- 40% Naphtha
- 8% Other oils
- 50% Natural gas.

The integration level is moderate, and most of the investment will be focused on adding and maintaining natural gas infrastructure and general upgrades of aging facilities.

3.5.1.5 Middle East

The Middle East is the region with the most activities in the area of integration or construction of integrated facilities. Its growing share of the global feed supply to petrochemicals is about 13% and has the following distribution:

- 30% Ethane
- 25% Naphtha
- 25% Other oils
- 20% Natural gas.

The current integration level is still considered moderate, but that will change soon. The investment activity in the area of integrated refining and petrochemical complexes is high as several world-class projects have been initiated in a push to make the Middle East the main processing hub for products feeding the chemical industry and other sectors.

3.5.1.6 Africa

The African countries could be considered the "sleeping beauties" as they are expected to be the next areas of economic growth. Certain countries such as Nigeria, Mozambique, and Cameroon will have an advantage over others due to the presence of natural resources. South Africa will most likely see a shift from its mainly coal-dominated industries to other feedstocks as an effort to support the decarbonization of the energy market. Africa only accounts for 2% of the global petrochemical feed supply with the following distribution:

- 15% Ethane
- 50% Natural gas
- 20% Coal
- 15% Other feedstocks.

Due to the lack of a well-developed oil industry, the integration level is low, and predicted investment in integration between refineries and petrochemical plants is also low.

3.5.1.7 Asia Pacific

The Asia Pacific region has benefited from the growing industries in China and India, and consumes 47% of the global petrochemical feed supply. The distribution of feeds is as follows:

- 5% Ethane
- 60% Naphtha
- 5% Other oils
- 5% Natural gas
- 25% Coal.

The integration level between refineries and petrochemical complexes is high, and the investment in this region is predicted to be high as the industries target to provide more affordable energy to the domestic market and cheap consumer goods to the global markets.

The short-term future of the global refinery and petrochemicals markets will most likely show a very simple picture that has North America and Western Europe sending refinery and intermediate products to the Middle East. The Middle East will develop into the main processing hub for petrochemicals to supply products to Asia to feed the growing bulk chemicals and specialty chemicals industry.

3.5.2 Impact on Product Value Chains

Vertical integration between refineries and petrochemical complexes is the most effective way to decrease the operational expenses and maintain competitive profit margins. Choosing the most effective operating model and the required level of integration across the value chain – for each asset and each region – will be crucial for improving margins and sustaining profitability in a volatile market. A holistic approach to determining market opportunities, technology options, and asset management strategies will be required to develop the most attractive configuration for an integrated complex in its respective economic environment.

Process technologies will continue to emerge and develop as integration studies begin to further clarify molecular pinch points in a complex just as heat integration has benefited over the last two decades with the development of heat transfer pinch analysis for energy savings and emission minimizations. The management of hydrogen and sulfur is a simple example where additional technologies have allowed the elimination of constraints and the improvement of the recycle value of various streams in any integrated complex.

Advanced application of information technology and data analytics for all the factors that are important to be modeled and adjusted with real-world input data on a dynamic basis will allow management to react to any changes in market conditions by optimizing the whole system. This includes the application of advanced and more sophisticated process simulation tools and the innovative modeling of existing process knowledge, and of improved planning and evaluation processes and software to feed targets to the optimization and control systems that run the integrated complex.

The integration of oil refining and petrochemicals will also require several business alignment processes as integration is one of the primary factors to remain among the leading players in the downstream industry and improve competitiveness on a global stage. This includes improving their respective organization. Operating the oil refining and petrochemical business under the same management with common objectives to optimize the assets in mature areas and develop projects in growth regions will allow companies to take advantage of the full potential of all markets. This means identifying synergies in the integrated organization and potentially reducing the weight of central functions by as much as 15%. Potential synergies may come from increased leverage vs. suppliers and partners – even in finance, we have been able to find some gains, merging maintenance functions, and merging finance and HR functions, and also from opportunities for hardware integration and energy optimization, reducing investment costs, operating costs, and carbon footprint of the integrated complex.

And let's not forget the potential benefit in integrated and optimized shutdown, maintenance, and turnaround practices.

Unless space constraints put limitations on the integration of the facilities, integration will add a lot of value and extra profit margin and make the whole complex more attractive by reducing the overall capital and operating costs for refinery and petrochemical complexes. Reducing the amount of capital required, driving higher returns on capital employed, improving yields, lowering utility consumption, and reducing operating costs per metric ton of product will determine which companies will survive and thrive in the long run.

3.5.2.1 Feedstock

Currently, most petrochemical facilities depend on the availability of feedstocks such as naphtha, ethane, or natural gas. The coal to chemicals route is also an important option for producers in some locations including China. The challenge for a standalone petrochemicals complex is the lack of control over the feedstock supply and the inability to change the adjustment of feedstock it receives. Defining the optimum configuration of an integrated complex is highly dependent on the type of crude oil available for processing in the refinery part, the desired product slate for intermediates and final products, and the overall flexibility required to deal with all potential market changes.

Once the refinery is decoupled from the energy and liquid fuel markets, it will allow operations to improve the properties for refinery products to increase the petrochemical yields that can be achieved. Synergistic effects from the increasing availability of feedstocks, the option to process and reprocess streams from both the refinery and the petrochemical plants, and lowering the costs per unit of final products will create financial benefits that couldn't be achieved as separate operations. Cost savings will mainly come from reduction of fixed cost, but in many cases, there will also be opportunities to reduce variable cost.

As is practice in some regions and facilities, the petrochemical feedstock base can be extended to non-conventional hydrocarbon sources such as heavy oils and residues. For example, the cracking furnace for a steam cracker can be designed to allow the cracking of streams such as unconverted oil from a conversion unit in the refinery such as the hydrocracker.

3.5.2.2 Offgas and Light Products

Refinery offgas is typically used to supplement the fuel gas system within the refinery. In an integrated complex, the offgas can be treated to be used as a feed to the steam cracker and its value captured at marginal natural gas price. This will require some modifications to plants that typically crack naphtha. These modifications can be executed in a way to also allow for the processing of LPG and some heavier refinery streams. More value can be added by using ethylene and propylene from the refinery in the production process of polymers such as polypropylene and high- or low-density PE.

3.5.2.3 Naphtha and Gasoline Blend Pool

The ability to adjust the feedstock to the steam cracker allows the integrated complex to adjust the yield structure to meet changing market demands. For example, processing light naphtha as a feedstock means ethylene production is higher compared to the full-range naphtha. In a steam cracker, light naphtha produces 4%–5% more ethylene compared to full-range naphtha. See Figure 2.6 for a comparison of expected yields from a steam cracker using different feedstocks.

Aromatic compounds from the refinery (BTX) are typically valued by their use as an octane-boosting additive to the gasoline blend pool. The same aromatic compounds in a petrochemical complex have not only a different usage, but also a different commercial value. An integrated complex can take advantage of the spread in value by directing aromatics to the market where they bring the most value at any given time. For example, the extraction of toluene from the gasoline pool for realizing a higher value as chemical stream for the Toluene diisocyanate (TDI) supply is only possible in a highly integrated complex. As a rule of thumb, the commercial value of an aromatic hydrocarbon in a petrochemical complex is typically higher than in a refinery.

In return, the integrated petrochemical plants could provide cheaper chemical components that the refinery might be able to use in economically justified production of motor fuels under the new quality and emission standards. By using the refinery low-value side products such as refinery light ends, LPG, and light naphtha in a steam cracker, and by using heavy naphtha in an aromatics plant as feedstock, these products can be upgraded to high-value petrochemicals. Petrochemical side products such as MTBE, alkylate from C_4 olefins, and heavy aromatics can be used in the refinery and can be blended into gas oil or the gasoline pool. Those low-value streams can be upgraded to high-value refinery fuels.

The price differential between natural gas and crude petrochemical feedstocks such as ethylene, propylene, butadiene, benzene, and paraxylene will be the key driver to consider natural gas as a refinery fuel for better integration with a petrochemical complex.

3.5.2.4 Heavier Products

As mentioned above, an integrated complex has the option to invest in cracking furnaces that can handle heavy feed streams such as unconverted oil from a hydrocracker and convert it into high-value olefins. In return, the heavy quench oil which is rich in aromatic compounds can be processed in a visbreaker or delayed coker in the refinery. As the refinery becomes more and more independent from the liquid fuel market, the feed quality for thermal and catalytic crackers in the refinery can be adjusted and optimized to maximize the yield of products that are valuable as feed streams to the petrochemical plants, for example, pentane and raffinates. By-products can be reprocessed to increase the output of more valuable products. Recycling of petrochemical by-product streams for fuel blending or reprocessing to petrochemical feedstock will allow a reduction in feedstock cost and an increase in product revenue.

3.5.2.5 Hydrogen

The refinery will still produce a significant amount of liquid fuels to meet the local and international demand from the transportation and power generation markets. Meeting the stringent and most likely becoming more restrictive quality requirements, for example, lower sulfur content in liquid fuels, will require the production of more hydrogen as the removal of the remaining sulfur-carrying molecules is becoming technically more challenging and more energy-intensive. Production of additional hydrogen would most likely occur via syngas production in a steam-methane reformer, which is costly and, in some cases, prohibitive to the economics of fuel production, providing hydrogen produced from petrochemical plants as a by-product. Optimization of the hydrogen network across the integrated complex and the recovery and reuse of hydrogen from the petrochemical side will allow a reduction in net hydrogen production costs, and, with this, a reduction in the overall cost of compliant liquid fuels production.

3.5.3 Impact on Storage, Logistics, and Infrastructure

The industry agrees that the integration of refining and petrochemicals will allow for the realization of significant cost savings across the complex. Cost savings will be realized in areas such as

- Storage, inventories, and product handling
  - For example, integrating the naphtha storage allows for optimizing capital employment whilst also reducing storage cost.
  - Reduced working capital.
- Common utilities and infrastructure, for example, buildings
  - Utilities and infrastructure can account for up to 40% of the total investment for a capital project.

- Energy integration and optimization
  - Higher overall energy efficiency.
  - Lower utility consumption.
- Centralized support services: security, storage, logistics, maintenance, finance, HR
  - Saving 10% in the supply chain of raw materials and intermediates can increase the profitability by 6%–8%.
  - Saving 10% in administrative cost can increase the profitability by 2%–4%.
- Transport and freight.

BIBLIOGRAPHY

Commissioned by the World Refining Association; Contributing authors: Kenan Yavuz, SOCAR Turkey; Nathalie Brunelle, TOTAL; Ivan Soucek, NIS; Gabor Kennessy, ORPIC; Global Report on Refining and Petrochemical Integration; 2014.

Crude Oil and Refining Market Outlook; IHS Markit; May 2018.

Energy Insights; Global Downstream Model; May 2019.

International Energy Agency (IEA); The Future of Petrochemicals; 2018.

International Gas Union (IGU); Marco Alvera, SNAM; Joe M. Kang, IGU; Alan Thomson, BCG; Barcelona, Spain; Global Gas Report 2019.

Is Peak Oil Demand in Sight? McKinsey & Company; Oil & Gas; June 2016.

Peering into Energy's Crystal Ball; Scott Nyquist, McKinsey & Company; Houston, Texas; July 2015.

Rethinking the Future of Plastics; McKinsey & Company; Houston, Texas ; February 2016.

S&P Global/Platts; Global petrochemical trends H1 2020; January 2020.

U.S. Energy Information Administration (EIA); Annual Energy Outlook (AEO) 2020 with projections to 2050; January 2020.

U.S. Energy Information Administration (EIA); Short-Term Energy Outlook (STEO); May 2020.

4 Technical View

Bringing the fundamentals of refining and petrochemicals and their respective outlooks on markets and economic impact together, this chapter will discuss the technical aspect of the integration between refining and petrochemicals. The integration can happen in different ways:

- Connecting existing assets that are in close enough vicinity via pipelines and implementing an integrated management and operating team. To some degree, this concept has been proven successful for many years in Gelsenkirchen, Germany. Under the name of Ruhr Oel GmbH, owned by Veba Oel AG and Petroleos de Venezuela, SA (PdVSA), the two refineries located in Gelsenkirchen, Scholven and Horst, were connected by multiple pipelines bridging the about 11 km of distance between the two locations. This allowed Ruhr Oel to operate the two sites as one big refining complex and take advantage of exchange of products between a great variety of processing units. The Scholven site is also host of two world-class steam crackers using naphtha from the refinery and unconverted oil from the hydrocracker as main feedstocks. The refinery also has aromatics plants to produce benzene, toluene, and xylene. The ethylene and propylene production went to another joint venture with facilities on site to produce polyethylene and polypropylene pellets. Other streams were sent either via pipeline or via truck to the nearby chemicals complex in Marl for use in the production of various chemicals. This is a great example for an integrated complex that developed and grew over many decades by adjusting its concept to changing market conditions.
- Adding petrochemical process units to an existing refinery or building a petrochemical plant in the vicinity of an existing refinery. To some degree, this has happened along the US Gulf Coast over the past 5 years as new steam cracker capacity and other downstream units have been built. It must be clarified that these capacities were added to take advantage of the abundance of ethane as feedstocks coming from the shale gas fields. However, this still opens the door for exploring and taking advantage of synergies with the neighboring refineries.
- Adding refining units to an existing petrochemical complex or building a refinery in the vicinity of an existing petrochemical complex. This scenario is highly unlikely for several reasons. First, if there was a source of crude oil available for processing and a market for refined products, chances are high that a refinery would have been built in that location a long time ago. It will be difficult to justify a new grassroots refinery, even with the benefits from the integration with petrochemicals. Having said that, a shift in global fuels

and petrochemicals demand may change the economic environment drastically enough to support this kind of investment in the future.
- Building a grassroots integrated refining and petrochemical complex. This is the route that is currently being followed in the Middle East and in China. These projects are world-class projects tailored toward establishing the Middle East as the main processing hub for petrochemicals to supply products to the Asian (mostly Chinese) chemical industry.

No matter what scenario applies, the technical options for integration are basically the same for all options. They are focused around securing the amount of feedstock to utilize the processing capacity to the maximum extent possible, adjusting feedstock quality to allow for optimized processing and maximized output of valuable products, and around taking advantage of opportunities to reduce operating expenses and become more competitive in the market.

4.1 FEEDSTOCKS AND PRODUCTS

The real challenge in utilizing the benefits of an integrated processing complex is the management of feed streams, intermediate product streams, and recycling or reprocessing streams. Knowing at any given time what is the best route for a product requires a clear strategy and detailed knowledge of the processing options that your complex has.

4.1.1 Crude Oil Selection

Crude selection for refineries is typically done based on or within the following constraints:

- Availability of crude oil for the specific location
    - Crude oils from local or close-by markets are preferred to keep transportation cost and logistics risk low.
    - Crude oils from more remote markets may be feasible depending on price and transportation cost and risk.
- Capability of assets to handle specific crude oils
    - Very heavy crude oils may be low in price, but they are typically high in sulfur content and other contaminants, and may be more corrosive than other crude oils.
    - If the refinery is not designed for heavy, sour crude oils, processing these crude oils will cause high corrosion rates and – consequently – failure of equipment and piping.
    - If the refinery was designed for heavy crude oils with a high content of residue, processing a light crude oil with a higher content of lighter materials may overload the piping and equipment handling the lighter components.
- Required product rates and qualities
    - The selected crude oils must be of such quality that they allow the refinery to produce the desired amount of its products at the specified qualities.

The solution to this problem is not trivial and requires the use of sophisticated tools. In most of the cases, the tool of choice is a Linear Programming Model (LP model) used for refinery and production planning, supporting the crude oil selection and supply process, the setting of operational targets, and the planning of product blending, inventory management, and sales (product loading etc.).

Linear programming or linear optimization is a method to calculate the best outcome in a mathematical model in which all requirements are represented by linear relationships. Best outcome can mean different things, for example, maximum profit, lowest cost, maximum production rate, best quality, or – as you probably have guessed already – a combination of these. In other words, the LP model is the algebraic description of the outcome to be achieved and of all constraints to be satisfied by the variables in the linear equations (see Figure 4.1).

Typical constraints for a refinery LP model would be

- List of available crude oils and their amounts
- Unit capacities and online factor
- Available storage volume for intermediates and final products
- Blending capacities
- Physical connections between units and tanks
- Availability of utilities and auxiliary units.

Once all unit operations and constraints are identified in the LP model, they will have to be described by a set of linear equations. The more equations are required to define a process operation, the more difficult it will be and the more time it will need to find a feasible solution.

The definition of a good set of linear equations that describes a unit operation well enough to give the refinery planner a good result is crucial and requires great care. Ideally, the base equations will be developed using known correlations and constraints. Then, a refinery wide test run would be conducted to develop a good and

- There are n variables in m constraints to be solved

$$Max / Min \quad Z = c_1x_1 + c_2x_2 + ... + c_nx_n$$

$S.t.$

$$a_{11}x_1 + a_{12}x_2 + ... + a_{1n}x_n \le / \ge / = b_1$$
$$a_{21}x_1 + a_{22}x_2 + ... + a_{2n}x_n \le / \ge / = b_2$$
$$.......... \quad .... \quad .... \quad .... \quad .... \quad ......$$
$$a_{m1}x_1 + a_{m2}x_2 + ... + a_{mn}x_n \le / \ge / = b_m$$

$$x_i \ge 0, \quad i = 1, 2, ..., n$$

FIGURE 4.1 Typical LP model structure

consolidated database for the planning engineer to fine-tune the factors in the correlation to match the specific process unit and its unique capacity and operating characteristics. These fine-tuned equations will be run against the "old" or established equations, and the output of the LP models will be compared against each other and against historic performance data to determine if the new, fine-tuned equations improve the predictive accuracy of the LP model.

Due to the availability of higher computing power and more sophisticated software, the trend will move toward more complex refinery models that more resemble a full process simulation model. These models have the advantage that they will carry the full stream composition and parameters through the complete refinery flowsheet, while the LP model only provides information on the components and parameters that are required for the linear equations as input. However, the more complex and detailed the planning model, the more time is required for maintenance and the more complex is the analysis of the results.

One of the outputs of the LP model is the blend of crude oils that the refinery should pursue on the market and the exact amount for each type of crude oil. In many cases, refineries have long-term contracts for crude oil delivery, and the swing of required amount going into processing can be achieved by managing crude oil inventories accordingly. As is valid for most optimization challenges, the lesser the variables, the easier to solve. A model for a refinery using 5–10 crude oils as potential feeds will solve faster than the same refinery model with 20 or more crude oils to choose from. The economic result however will be better as the crude oil blend can be adjusted much better to fit the production and quality requirements.

Crude oils are defined by structured data sets called assays. Each crude oil gets tested in an accredited laboratory, and the test results are transferred into the structured form of the assay. The provision of a crude oil assay is a prerequisite for bringing the crude oil to market. And the extensive amount of data in the assay allows companies such as traders and refinery operators to analyze the compatibility of the crude oil with their current mix and the potential product yields and qualities from processing that specific crude oil. Crude oil qualities change over time as the quality of a crude oil coming from a specific set of oil wells will decline the longer the well operates. In that case, the oil must be tested again, and the assay will have to be revised accordingly. Some oil fields are so big that oil wells at different ends of the same oil field may produce a slightly different quality of the same crude oil. For that reason, you may have two slightly different assays for the same crude oil. And to complete the picture, some exploration companies blend different crude oils and market a crude oil blend, for example, REB = Russian Export Blend. The assay for a crude oil blend will change constantly as the amount of specific crude oils in the blend will change from delivery to delivery. In these cases, the general crude oil blend assay will have to show data ranges for planning purposes, and the customer may demand an updated, specific assay for the crude oil blend with each delivery.

Based on the quick introduction into refinery planning and crude oil selection, now imagine the refinery being integrated with a petrochemical complex. The integration will require changes to the LP model, for example:

- New process units may need to be added to reflect the added processing capabilities: aromatics unit, steam cracker, and other downstream petrochemicals units.
- New feed and product streams may need to be defined: naphtha may have to be defined as steam cracker feed, olefins from the FCC Unit, or any stream from a petrochemical unit that will be a new feed stream to a refinery unit or for blending.
- The price scenario will have to be modified to reflect the new value for intermediates and final products that have been developed for the integrated complex.

Once all the required changes have been made, the LP model (or any other planning model) will optimize the operational plan for the integrated complex and the selection of crude oils to be processed in the refinery part will change based on the requirements and constraints of the integrated complex model. For a complex with low level of integration, the selection of crude oils might not change much. The model will most likely select the same or a similar crude oil mix and change the individual flow rates to meet the new requirements. For a petrochemical refinery, the changes will be more drastic as different, new crude oils may be added to the mix that were unattractive before the integration but became attractive due to the changes in the objectives of the LP model. On the other hand, some crude oils that were a good fit for the refinery itself may be eliminated from the new crude oil mix for the same reason mentioned above. There are several potential drivers that will influence the crude oil selection:

- Content of sulfur and other contaminants: while the refinery may have a higher tolerance for sulfur, nitrogen, metals, or other contaminants, the integrated complex may have to use more rigorous constraints.
- Yield of certain products: based on the demand for certain products, the model may favor crude oils that deliver a higher yield of straight-run products in the desired boiling range (for example, naphtha) or a higher yield of certain intermediates to improve the yields in one or several of the refinery's conversion units.
- Price: if the value generation from products coming from the integrated complex increases, some higher priced crude oils that were ignored in previous LP model runs may become viable as feed to the refinery, if in parallel they allow optimization of product yields and/or qualities.

Based on the availability of suitable crude oils for petrochemical manufacturing such as Arabian Light crude oil, capital investment will flow into the Middle East and Asia to build world-class integrated refining and petrochemical complexes that will serve local and global markets, for example, for olefins, aromatics, glycols, and polymers. One of the major oil companies has technology to directly crack crude oil and eliminate the refining part of the value chain. This may be beneficial in some cases, but it also takes away a lot of the flexibility that an integrated refining and petrochemical complex offers.

Since we will cover the role of other feedstocks in a separate section, we can now have a closer look at the product value chains of an integrated complex.

4.1.2 Refinery Product Value Chains

The discussion of product value chains in an integrated refinery and petrochemical complex can only give a generic view on all options that are technically available. The actual realization of these options will depend on the specific configuration of each complex, and the regional markets these facilities will operate in. For example, increasing the integration between the fluid catalytic cracking (FCC) unit and petrochemicals may be critical to the economically feasible operation of this unit if connected to a declining gasoline market. In another location where the gasoline market is still strong, the FFC unit may continue to primarily provide naphtha for the gasoline blend pool. Another significant factor will be the environmental regulations the complex has to comply with, as some markets may adopt more stringent limits on contaminants such as sulfur faster and earlier than others.

4.1.2.1 Refinery Gases

For the purpose of organizing the information we decided to look at product value chains starting from light to heavy products and going from refinery to petrochemicals. Refinery gases are gaseous product streams coming from the following sources:

- Light components as part of the crude oil blend processed and separated in the crude oil distillation columns – the atmospheric and the vacuum distillation units.
- Light components created by cracking of heavier components in conversion units – thermal or catalytic cracking units.

The main components of refinery gases sorted by carbon number are

- C_1 – methane
- C_2 – ethane, ethylene
- C_3 – propane, propylene
- C_4 – normal butane, isobutane, and others.

In a normal refinery configuration, methane and some of the ethane/ethylene mix would be utilized to supplement the refinery fuel gas system. Gases such as propane and butane can be sold in liquid form as liquified petroleum gas (LPG) for cooking and as fuel. In an integrated complex, there are more options for utilization of these products. LPG can also be used as a feed to naphtha and LPG crackers. The C_4 hydrocarbon mix coming from the cracker unit can be processed further to butadiene or other products.

4.1.2.1.1 Methane

Utilizing methane-rich gas as fuel gas may still be a viable option for an integrated complex as it will allow reducing the demand for natural gas or other fuels to run the complex.

The added option in an integrated complex is the conversion of methane to methanol. Methanol can be used on the petrochemical side for multiple purposes as described in Section 4.1.3.5.

4.1.2.1.2 Ethane

Ethane is one of the two dominant feeds being processed in steam crackers, with the other feed being naphtha. In most cases, defining the value of ethane as steam cracker feed will be more favorable than the value as fuel gas supplement. In an integrated complex, ethane will most likely be routed to the steam cracker to maximize the profit of the complex. Using ethane as steam cracker feed will drive the production toward ethylene, a petrochemical product which we will discuss in Section 4.1.3.1. This is the most favorable in complexes that put a higher value on ethylene than on propylene and want to maximize ethylene production. In general, ethylene is considered the more valuable product for petrochemical manufacturing, with propylene ranking as number two.

4.1.2.1.3 Propane and LPG

Propane can either be sold as liquid fuel (LPG) or separated from the C_4 hydrocarbons and used as a separate product. One of the options to feed propane and LPG to the petrochemical section of the complex is the use of the stream as feed to the steam cracker. The propane portion of the feed will convert to ethylene and propylene, while the heavier components of the LPG will convert to propylene and a stream that is called pyrolysis gasoline, or short pygas. Pygas is produced primarily in naphtha-based steam crackers and is a naphtha-range product with a high aromatics content. We will discuss the use of pygas in Section 4.1.3.4.

If propane is separated from the LPG, it can be fed into a propane dehydrogenation unit. Propane dehydrogenation (PDH) is a selective catalytic process that removes hydrogen from the stream and converts propane into propylene without the need to send it through a steam cracker. This frees up steam cracker capacity to process ethane and/or naphtha. The process also has a big advantage; that is, it creates hydrogen as a by-product. Hydrogen is in high demand on the refinery side as it is required for the desulfurization of fuels and other products. Manufacturing on-demand hydrogen via steam-methane reforming (SMR) and purification via pressure swing absorption (PSA) are capital- and energy-intensive. Providing hydrogen as a by-product from the petrochemical side represents great value to the refining side and will make the production of fuels more competitive due to reduced manufacturing cost.

4.1.2.2 Naphtha

Naphtha is one of the component groups that forms crude oil and will be separated from the other product groups in the crude oil distillation section of the refinery. This product is also referred to as virgin or straight-run naphtha. Naphtha is also produced as a product from cracking or conversion units that convert heavier, less valuable products into lighter, more valuable streams. The naphtha stream from a conversion unit is also referred to as cracked naphtha or named after the unit it is produced in, for example, Coker naphtha or FCC naphtha.

Naphtha can be separated further into streams called light naphtha and heavy Naphtha, depending on where the stream will be used and what quality requirements come with the destination.

Naphtha can be used neat in the gasoline blending pool, but in most cases, it needs to be "upgraded" by boosting the octane number. The most common process is the continuous catalytic reformer (CCR) or Platformer. This process is designed to increase the octane number of a low-octane naphtha stream as a main component for the gasoline blending pool. This product is called Reformate or Platformate.

In an integrated complex, the alternative outlet for naphtha is the use as feed to the steam cracker to produce ethylene, propylene, C_4 olefins, and pygas. Reformate can be used in the pygas pool to produce aromatics.

4.1.2.3 Other Refinery Products

All other refinery products that are heavier than naphtha play no significant role in the petrochemical sector. Jet/kerosene, diesel, gasoil/fuel oil, and asphalt will follow their traditional value chains and will not be impacted by the integration with petrochemicals.

The only exception may be the unconverted gasoil stream from a hydrocracker or a similar gasoil stream, which can be used as heavy feed to specially designed steam cracking furnaces.

4.1.2.4 Aromatics (BTX) from Refinery Units

Some refinery operations have an aromatics section included in their process scheme that allows the separation of aromatics-rich naphtha streams into its components: benzene, toluene, and xylene for further use in the petrochemical value chain (see Section 4.1.3.4).

4.1.2.5 Olefins from Refinery Units

Refinery gas coming from a conversion unit, predominantly the FCC unit, contains a noticeable quantity of olefins such as ethylene and propylene that – in some cases – is high enough to consider separation of the olefins and use in the petrochemical manufacturing value chain. This route becomes more attractive in an integrated complex as the outlets for ethylene and propylene are available to the refinery side of the complex. The value of olefins in the petrochemical manufacturing chain will be higher than their value as fuel gas at natural gas price.

4.1.3 Petrochemical Product Value Chains

The description of the refinery product value chains is quite short as the value increase comes from the opportunity to process these streams in the petrochemical units, and their use is described in more detail in this section covering the petrochemical product value chains.

4.1.3.1 Ethylene

Ethylene is the main product coming from a steam cracker, especially when the cracker feed is ethane vs. any of the other potential feed streams. Ethylene is the base stock for multiple manufacturing processes, so its value chain spreads out over a wide range of products.

- Polyethylene (PE)
  - Low-density PE (LDPE) – used in food packaging
  - Low low-density PE (LLDPE) – used in food packaging
  - High-density PE (HDPE) – used in food packaging.
- Ethylene oxide (EO)
  - Ethanolamines
    - Monolethanolamine (MEA) – used in detergents
    - Diethanolamine (DEA) – used in agrochemicals
    - Triethanolamine (TEA) – used in detergents.
  - Ethylene glycol/mono ethylene glycol – used as engine coolant or in solvents
    - Polyesters – used in food packaging and fibers.
  - Ethylene glycol ethers – used in solvents
    - Ethylene glycol ethers acetates – used in solvents.
  - Diethylene glycol – used in solvents
  - Solvents are used to manufacture a variety of final products such as coatings, adhesives, inks, detergents, and pharmaceuticals. Solvents are also used in chemical synthesis processes.
- Ethylbenzene
  - Styrene
    - Acrylonitrile butadiene styrene (ABS) – used for automotive parts and other formed plastics (e.g., luggage)
    - Styrene acrylonitrile resin (SAN) – used for automotive parts and other formed plastics (e.g., luggage)
    - Unsaturated polyesters – used in glass reinforced plastics
    - Polystyrene plastics – used in consumer electronics
    - Styrene-butadiene rubber (SBR)/synthetic rubber – used for tires and in the automotive and equipment industry.
- Ethylene dichloride
  - Vinyl chloride monomer
    - Polyvinyl chloride (PVC) plastics – used in construction and for medical equipment.
- Ethyl alcohol
  - Ethyl acrylate – used in footwear and tires
    - Acrylate elastomers – used in footwear and tires.
- Synthetic rubbers – used in footwear and tires
  - Ethyl acetate – used in solvents
  - Ethyl amines – used in the paper production.
- Acetaldehyde
  - Acetic acid
    - Vinyl acetate monomer – used in adhesives
- Vinyl acetate plastics – used in electric components and electronic appliances.

4.1.3.2 Propylene

Propylene is another product from the steam cracker and – such as ethylene – has a variety of uses in the olefin-based production of petrochemicals and chemicals.

- Polypropylene (PP) – used in food packaging.
- Propylene oxide (PO)
    - Propylene glycol / mono propylene glycol – used in the marine industry, in bath ware, as engine coolant, as solvent, in food processing, in pharmaceuticals, and in cosmetics
        - Polyester resins – used in solvents.
    - Propylene glycols (Di-, Tri-, …) – used in solvents
    - Propylene glycol ethers – used in solvents
        - Propylene glycol ether acetates – used in solvents.
    - Polyols
        - Polyurethane – used in sportswear.
- Cumene
    - Phenol
        - Phenolic resins – used in furniture
        - Bisphenol A – used for epoxy resin.
- Polycarbonate – used in construction
    - Acetone – used in bisphenol A and in solvents
        - Methyl methacrylate (MMA).
            - Polymethyl methacrylate (PMMA) – used in unbreakable glass
- Acrylonitrile (ACN)
    - Acrylonitrile butadiene styrene (ABS) – used as plastics for automotive parts and other formed plastics (e.g., luggage)
    - Acrylonitrile butadiene rubber (NBR) – used as plastics in the automotive industry
    - Styrene acrylonitrile resin (SAN) – used for automotive parts and other formed plastics (e.g., luggage)
- Acrylic fibers – used in the fabrication of textiles.
- Acrylic acid (AA)
    - Polyacrylate
        - Superabsorbents – used in disposable diapers and nappies.
- Butyraldehyde
    - Ethyl hexanol
        - Plasticizers – used in kitchen appliances.
    - n-Butanol – used in acrylic paints
        - n-Butylacetate – used in solvents.
    - Isobutanol.
- Isobutyl acetate – used in solvents
    - Acrylic acid (AA)
        - Acrylic esters – used in paints, coatings, textiles and adhesives.
- Isopropanol
    - Acetone
        - Methyl methacrylate (MMA).

- Polymethyl methacrylate (PMMA) – used in unbreakable glass
- Isophorone – used in solvents.
- Isopropyl acetate – used in solvents.

4.1.3.3 C_4 Hydrocarbons

The role of C_4 hydrocarbons is not as diverse as for the lighter products, but there are multiple paths for processing this stream in the petrochemicals sector.

- N-Butenes
  - Higher olefins – used in detergents and agrochemicals.
- Butadiene
  - Nitrite-butadiene rubber (NBR) – used in the automotive industry
  - Styrene-butadiene rubber (SBR) / Synthetic rubber – used in the automotive and equipment industry.
- Isobutylene
  - Polyisobutylene – used in engine lubricants
  - Butyl rubbers – used in food products such as chewing gum
  - Methyl tert-butyl ether (MTBE) – used as additive in the gasoline pool.

4.1.3.4 Pyrolysis Gasoline (Pygas)

The pygas stream is the main carrier of aromatics that have a variety of uses in the petrochemical industry. Based on the names of the aromatic components, benzene, toluene, and xylene, the stream itself and the associated separation unit are often referred to as BTX stream and BTX unit. Of the three aromatic components, benzene is the most versatile aromatic.

- Benzene
  - Ethylbenzene
    - Styrene.
      - Polystyrene plastics (PS) – used in food packaging
      - Styrene-butadiene rubber (SBR) / Synthetic rubber – used in the automotive and equipment industry
      - Acrylonitrile butadiene styrene (ABS) – used for automotive parts and other formed plastics (e.g., luggage)
      - Styrene acrylonitrile resin (SAN) – used for automotive parts and other formed plastics (e.g., luggage)
      - Unsaturated polyester
  - Cumene
    - Phenol.
      - Phenolic resins – used in furniture
      - Bisphenol A
        - o Polycarbonate – used in CDs
        - o Epoxy resin
    - Acetone – used as solvent.
      - Bisphenol A
        - o Polycarbonate – used in CDs
        - o Epoxy resin.

        - Methyl methacrylate (MMA)
            - o Polymethyl methacrylate (PMMA) – used in unbreakable glass
    - Cyclohexane
        - Nylon – used on sports equipment and textile fibers.
    - Methyl diphenyl diisocyanate (MDI)
        - Polyurethane – used in sportswear.
    - Alkylbenzene – used in surfactants for detergents.
- Toluene – used in solvents
    - Toluene diisocyanate (TDI)
        - Polyurethane – used in sportswear.
- Xylenes
    - Paraxylene
        - Polyester – used in the textile industry.
    - Orthoxylene
        - Plasticizers.
            - Flexible PVC – used in the building and construction industry.

4.1.3.5 Methanol

The alternative route to using methane as fuel gas is the conversion to methanol and its use in the petrochemical manufacturing process.

- Methanol
    - Methyl methacrylate (MMA)
        - Polymethyl methacrylate (PMMA) – used in unbreakable glass.
    - Formaldehyde
        - Phenolic resins – used in thermal insulation
        - Polyurethane – used in home furnishings and sports equipment.
    - Methyl tert-butyl ether (MTBE) – used as additive in the gasoline pool
    - Other chemical intermediates
    - Use of methanol as fuel or as precursor to fuel (in discussion).

4.1.3.6 Other Petrochemical Products

The list of petrochemical products as provided in the sections above is not a complete list of all possible products that can be manufactured from the seven base petrochemicals. For example, polystyrene or PS can be further altered into expanded polystyrene or EPS, which in some countries is known as Styrofoam. Styrofoam is used to great extent in the packaging industry and as insulation material. There are several examples for specialty chemicals and other products that can be produced from petrochemicals, but we believe we have covered the most significant ones that would apply to the concept of an integrated refining and petrochemical complex.

4.1.4 Selection of Integration Options

The question of what level of integration is best for each complex strongly depends on the basis the complex is built upon. Not all projects can be developed from the same starting point. Consequently, the methodology that needs to be applied to the selection of what value chains the complex can take advantage of will differ with each case. Here are the four basic cases that can be defined:

- Grassroots complex: new refinery and new petrochemical units; no existing facilities; all processing routes can be selected
- Existing refinery: new petrochemical units; existing facilities; selection of process routes might be constrained
- Existing petrochemical units: new refinery units; selection of process routes will be constrained
- Existing integrated complex with low level of integration: achieve higher level of integration; main process routes already selected.

Before we look at each of the cases, here is a reminder of what tools we have at our hands to make any decisions with regard to the level of integration and what process routes are the best to select. First, we must look at the market analysis, regional and global, to identify the demand and supply balance and the 5-, 10-, and 20-year forecast for all petrochemicals that we are interested in. This is a very critical step in the evaluation and development of the integration project for the following reasons:

- Any investment decision of this order of magnitude has to be based on a solid determination that there is or there will be a long-term demand for the product or products that the new units will produce and that the product or products can be brought to market in a competitive manner.
- Infrastructure needs to be in place or must be installed economically to bring the product or products to the target market or markets.
- The determination what product or products are in high demand will impact the selection of processing routes required to manufacture the desired product or products. In most cases, this will decide the feasibility of the project.

Many project ideas and industrial developments fail because they apply a new and attractive technology to produce a high-value product, but they have no market for the product or are not competitive in the market they need to sell in. Spending the right amount of time and money on the market analysis and on developing a long-term sales strategy is a crucial initial step in a successful project execution.

Once the market analysis is done and the product or products that can be sold are identified, it is time to determine the manufacturing route and the required feedstocks. In some cases, this needs to be an iterative process between the technical team and the commercial/marketing team. Should the market analysis result in the detailed specification of a well-defined product, there is a high probability that there is only one process option to choose and that the only selection process to be applied will be between multiple licensors for this process. In other cases, the market analysis will determine the demand for a product group or a product with a certain range in acceptable quality. In these cases, the technical team and the commercial team must agree on the exact product specification they want to use for developing the technical concept of the project. Here are a few examples for a better understanding of this decision process.

Example 1: Synthetic Rubber

In this example, the market analysis for the target region of your integrated complex has determined a strong long-term demand for synthetic rubber to support the growth of an industry that needs to find ways to replace its current supply of natural rubber with synthetic rubber. One decision is easy: you need processing capabilities for your C_4 hydrocarbon streams to produce butadiene. But the question is what synthetic rubber to produce.

- Option 1: Styrene-butadiene rubber (SBR) will require process units to produce styrene. The feed streams needed to produce styrene are ethylene and benzene which will be used to produce ethylbenzene. Ethylbenzene will then be converted to styrene. It also needs to be determined if ethylbenzene is produced via dehydrogenation (delivering excess hydrogen to the refinery) or via an ethylbenzene hydroperoxide process.
- Option 2: Acrylonitrile butadiene rubber (NBR) will require propylene and ammonia to produce acrylonitrile.

These two options require different units and different feedstocks, so it is critical to select the right value chain for the integrated complex to ensure that the market demand and product specification needs can be satisfied.

NBR and natural rubber are very similar in molecular structure up to the point that the elastomers may seem like identical twins. They both possess superior physical properties including high tensile strength, low compression set, and good abrasion resistance. NBR and natural rubber even share the same resistant properties to UV rays, ozone, and excessive heat. However, NBR has one edge over natural rubber which is its resistance to oils, grease, and other fuels.

SBR is the best synthetic elastomer to use in dynamic and abrasive conditions. SBR will not easily wear away as a result of abrasions. In addition, SBR is one of the lowest cost rubber products. NBR has a moderate resistance to abrasions in addition to a higher tensile strength and a lower compression set. The key feature that distinguishes the two is that NBR is an oil-resistant nitrile elastomer.

If the desired rubber product or products will come into contact with petroleum by-products and will see constant abrasion, then NBR is the perfect product for the task. In all other cases, SBR is the product of choice.

Example 2: Solvent

The term "solvent" describes a substance with the capability to dissolve a solute to form a solution. The most common solvent is water which has the capability to dissolve polar molecules such as ions and proteins, a process that is used in the human body and all other living organisms. In case of solvents from petrochemicals, the exact use of the solvent determines which type is best suited for the application. Examples are

- Dry cleaning: tetrachloroethylene
- Adhesives: ethyl acetate, methyl acetate, acetone
- Cosmetics/perfume: ethanol
- Paint thinner: toluene.

The definition of the final use of the solvent will define which product will be required and what production processes need to be selected. This will also determine the required feed streams for the process.

Once these decisions have been made, the technical feasibility must be determined. In many cases, this will require input from technology licensors. The process to obtain technical data from licensors includes the following steps:

- Execution of a Non-Disclosure Agreement (to protect the intellectual property/license).
- Development and submission of a Performance Specification for the process.
- Responding to all licensor questions.
- Once the licensor information is received, the technical team can apply several tools to analyze and compare the available options.
- Information can be compiled, presented in a spreadsheet such as MS Excel or similar. Formulas and graphic tools can be used to create simple flowsheets and a simple material balance.
- A LP model can be either created or modified to confirm the economics of the new operations scheme.
- A detailed simulation model can be created for each case. This might be difficult as the package from the licensors may not have enough information to get a process simulation converged. However, a simulation model would be extremely helpful as it would allow for additional tasks to be completed as follows:
  - Preliminary sizing of all major equipment
  - Use of equipment sizes for initial estimate of the Total Installed Cost (TIC) for each case.

The application of these tools in support of the selection of integration options will vary for each of the four base cases we defined at the beginning of this section. However, the hierarchy for the economic drivers for integration shall be the same for all cases:

1. The highest priority is given to the value of the petrochemical products. Any objective around quantities and revenue from petrochemical products will be optimized first.
2. Secondary priority is given to fuels such as gasoline, jet fuel, diesel, and low-sulfur bunker fuel.
3. All other products have lower priority.

And as usual, there will be exceptions based on very specific demands from a geographically or economically isolated market sector.

4.1.4.1 Grassroots Refinery/Petrochemical Complex or Petrochemical Refinery

Starting from a clean sheet allows the project team to consider all options available within the framework given by the market demands and the available capital. World-class projects are currently planned only for the Middle East and Asia to support the growing markets in China, India, and the Philippines. These complexes also don't have geographical constraints as the high volume of production will allow them to be competitive even in remote markets, especially if there is a shortage of supply and no local competition. The planning and selection of processing routes is very complex and will require time and effort to allow for the appropriate due diligence. It is unlikely that a process simulation for the full complex will be developed, but parts of the complex will be simulated separately to create a complete heat and material balance for the assessment of product yields and qualities. The economic analysis will be supported by an LP model. Due to its complexity, the time and computing power required will have to be accounted for in the schedule for the selection process.

The capacities of all petrochemical units selected will determine the size of the steam cracker that will supply the feed streams to these units. The required yields of the steam cracker for ethylene, propylene, and C_4 hydrocarbons and pyrolysis gasoline determine the blend of feed streams such as ethane, naphtha, or LPG that need to be provided to the steam cracker. Unless the concept includes the import of feed material for the steam cracker, the feed requirements determine the capacity of the refinery units that supply these feeds for steam cracking.

4.1.4.2 Existing Refinery

To add petrochemical units to an existing refinery has a few constraints that limit the extent of the integration between the existing and the new units. It is logical to evaluate the existing product streams to identify the quantity of feed material for the steam cracker and the petrochemical units. Based on the results of the market study, the yields and capacities for petrochemicals can be compared to the desired product slate. The comparison will lead to a gap analysis that will show where the products from the refinery fall short in delivering what the petrochemical side needs. Any gaps can either be filled by import of feed material to the steam cracker, by import of intermediate products, or by investing in modifications and debottlenecking of the refinery to make its product yields fit the petrochemical side. Products that are rerouted from fuel production to petrochemicals will reduce the volume of fuels produced by the refinery.

4.1.4.3 Existing Petrochemical Units

The first step in this case is to check the market data vs. the output of the petrochemical units and identify any gaps that need to be filled. This may lead to additional capacities or to the addition of new units to open the door to new markets with higher potential for growth. Once this analysis is completed and the scope of modifications or additions to the petrochemical side is complete, the material balance around the petrochemical units will determine the amount and quality of feed that needs to be provided to the steam cracker. So, the quantities of naphtha and ethane and other refinery products will determine the selection of crude oils to be processed, the size of the crude oil distillation column, and the capacity and design of the conversion units to be included in the refinery flow scheme. Once the refinery output is matched up with the petrochemical complex, the production of liquid fuels and other refinery products will come out of the refinery's material balance and will need to be matched with the appropriate markets in the region.

4.1.4.3 Existing Integrated Complex

In this case, all tools should be in place and the scope of modifications or additions will come from the gap analysis using the market data provided. All identified gaps and bottlenecks can be addressed either by modifications to the existing assets, by adding the respective capacity in new units, or in some cases by optimizing the routing of streams and giving up capacity for production of lower value products to gain capacity for the production of higher value products. Another option to look for is the addition of infrastructure such as storage tanks, piping, and pumps to increase the level of integration between the refinery side and the petrochemical side. This will give the complex more flexibility in exchanging products and increase the amount of material coming from the refinery that will be used in the manufacturing process of petrochemicals.

Once the production routes for the refinery and petrochemical products have been determined, as a last optimization step some of the by-products from the petrochemical side can be recycled back into the refinery and used as blend stock for the fuel production or as feed material to the refinery's conversion units.

4.1.5 Role of Alternative Feedstocks

As shown in the economic view on current markets and market trends, specific developments in certain regions and economies will drive the selection of feedstocks to the steam crackers. To keep up with the growth rate of their economies and to keep prices competitive and affordable, some Asian companies – especially Chinese companies – will base their petrochemicals production on coal. Due to a stagnant fuel market, European companies utilize naphtha as their main feed stream to produce petrochemicals. And thanks to the shale gas revolution on North America, US Gulf Coast companies selected ethane as their primary feed stream for steam cracking.

On the technical side, the selection of the feed stream or blend of feed streams determines the yields of the steam cracker and, consequently, the quantities of intermediates that can be processed into final products for consumption.

4.1.5.1 Coal

Coal as a solid fossil fuel can't be processed in a petrochemical facility and needs to go through a gasification step that converts coal into gaseous and liquid products that can be used as feed streams to the petrochemical complex. This gasification step makes the use of coal capital intensive and raises concerns due to the carbon emissions associated with the process. The main products from the coal gasification and treatment of products that are typically used as feed to a petrochemical facility are naphtha and methanol. Consequently, the petrochemical facility could be designed for any of the processing routes. Naphtha is converted into ethylene, propylene, C4 hydrocarbons, and pyrolysis gasoline, and by adding methanol to the mix, the integrated complex has all five intermediate streams that can feed the whole variety of processes as described in the value chain section. If the economic model for the complex works and any environmental concerns can be mitigated, coal provides for a feed source with a high degree of flexibility.

4.1.5.2 Natural Gas

Natural gas as feed to a petrochemical unit provides the following molecules:

- Methane (>90% of natural gas) – conversion to methanol and its value chain
- Ethane – feed to the steam cracker for ethylene
- LPG (propane and butane) – feed to the steam cracker for ethylene, propylene, and C4's
- Natural gas liquids (naphtha) – feed to the steam cracker.

In regions that consider natural gas as feedstock, it is an inexpensive commodity and it provides decent yields of ethylene. Since over 90% of the natural gas is methane, it doesn't provide for a lot of flexibility in the selection processing routes for petrochemicals.

4.1.5.3 Biomass and Renewables

The current efforts in research and technology development involving biomass and renewable sources focus on the decarbonization of the fuels and energy markets. The growth potential in these areas is immense, and the most value of biofuels and bioenergy is currently seen in the replacement of fossil fuels and fossil-based energy. Experts also see biofuels and bioenergy as one of the measures that will help lowering global carbon emissions and slowing down the negative effects of these emissions on the global climate. Very little effort is put into the development of technologies that will allow the use of biomass and renewables in the petrochemical sector. Here are two examples of technologies that are being considered:

- Fermentation of biomass is currently used to produce bioethanol for the fuel market. The same technology can be utilized to produce biomethanol as base feed for petrochemicals such as polyurethane or polymethyl methacrylate (PMMA).
- Virent Inc. – in cooperation with BP and Johnson Matthey – developed a process called Bioforming® which mimics the catalytic reforming of petroleum naphtha and converts biomass into a reformate stream that can be used to produce bio-paraxylene (bio PX), which can be processed further to bio-purified terephthalic acid (bio PTA) or renewable polyester.

Once biomass and other renewable energy sources have been established in the liquid fuels value chains, they will continue to develop into technologies that will feed the petrochemical sector and replace carbon-intensive fossil feed sources.

4.2 ENERGY EFFICIENCY

One of the big drivers for the integration between the refinery side and the petrochemical side is the gain in competitiveness on the market and the chance to beat the competitors by operational excellence and superior performance of the integrated complex. And one of the important pieces to that puzzle is the uplift in energy efficiency compared to standalone operations. Any reduction in the use of energy or any improvement in the efficiency of the use of energy will allow a reduction of variable cost of production and an increase in production margin.

The first step that needs to be undertaken is the understanding of where energy is used within the complex, what it is used for, and how much is used. The numbers presented in this chapter are average numbers that were developed based on information from several industry reports. These numbers may not completely accurately describe any complex, but they give a good understanding of energy usage within the two sides of the integrated complex and – as a result – a good indication where the most impact can be made by implementing improvements.

Although in literature there are references to what a typical conversion refinery flow scheme would look like, it is highly likely that each refinery is a little different and has certain process units that others don't have or has a combination of units that is rather unique than common. Table 4.1 shows the typical range of energy consumption (all sources of energy combined) as a percentage of the total energy consumption of that specific, highly complex refinery configuration.

The numbers in Table 4.1 represent an example for a high complexity conversion refinery with a fluid catalytic cracker and a CCR for gasoline production, a hydrocracker for diesel production, and a visbreaker/delayed coker section for residue cracking and asphalt production. For clarification, all energy coming from the refinery utilities such as steam or cooling electricity has been assigned to each process unit based on its consumption of the respective energy form. Consequently, reduction in energy consumption in one or several process units will impact the overall load of the utility units and may reduce their efficiency. For example, if steam consumption is reduced drastically and goes past the minimum load point at which the steam boiler can be operated, it will most likely keep operating at its safe minimum load and excess steam will be vented to the atmosphere. This is an extreme example, but

TABLE 4.1
Unit Specific Energy Consumption as % of Total Refinery Energy Consumption

Process Unit	Specific Use as % of Total Refinery Consumption
Desalting	0.1–0.5
Crude distillation unit	30–35
Vacuum distillation unit	10–15
Gas separation unit	1–5
Visbreaking unit	1–5
Delayed coking unit	3–6
Fluid catalytic cracking unit	5–10
Hydrocracking unit	10–15
Continuous catalytic reforming unit	5–10
Hydrotreating unit	15–20
Alkylation unit	1–5
Isomerization unit	0.1–0.5

it highlights the need to evaluate and optimize the energy system as a whole and look at every aspect of the energy networks within the refinery or integrated complex.

As can be seen from the number ranges, the most likely candidates for energy optimization programs are the units with the highest energy consumption rates, which are the crude distillation unit, the hydrotreating units, the vacuum distillation, and the hydrocracking unit. In simple words, producing low sulfur or even ultra-low sulfur diesel requires the most energy-intensive processes in a refinery. In many cases, operating companies do not work through energy conservation programs by unit but by consumer type. For example, there are a lot of things that can be improved around fired heaters that will minimize the use of fuel gas (or any other fuel) and create savings across the whole refinery. Similar programs are being executed for pumps and compressors, for air coolers or for distillation columns (see Section 4.2.1). There are reports available that list numerous potential measures that can be taken for these systems and types of equipment that will reduce the overall energy consumption of the facility. Table 4.2 shows the distribution of energy consumed in a typical refinery by type.

In our example refinery, about 60%–65% of all energy consumed is fuel to fired heaters and other heat generation systems. Fuel consists mostly of fuel gas, but can also be fuel oil, especially in older facilities, or even other types of fuel. Steam generation and electricity share the rest of the energy consumption. In summary, executing an energy survey and optimization program in a typical, high-complexity conversion refinery will most likely yield the following results:

- Most savings will be centered around fuel (most likely fuel gas) consumption.
- Some savings will be achieved for electricity and steam consumption. And yes, reduction in steam consumption should also result in fuel savings.
- The most likely beneficiary of the program will be ultra-low sulfur diesel, but gasoline and other products will also benefit from reduction in variable operating expenses.

Shifting the focus to the petrochemical side of the facility, we can look at similar data to help us orient and determine areas of high probability for improvement. Table 4.3 shows the energy use in different sectors of the petrochemicals and chemicals industry as percentage of the total energy use.

With a share of 15%–20% of the total energy consumption, petrochemicals and plastics/resins are each on the lower end of the spectrum. Specialty chemicals and other organic chemistry are leading the industry in energy consumption with a share of 30%–35% each. Almost half of the energy comes from natural gas, followed by renewable energy sources, electricity, and the internal use of by-products

TABLE 4.2
Refinery Energy Consumption by Type

Steam generation	15%–20%
Fuel	60%–65%
Electricity	15%–20%

TABLE 4.3
Energy Use by Industry Sector Petrochemicals and Chemicals

Industry Sector	Energy Use (%)
Petrochemicals	15–20
Other organics	30–35
Plastics and resins	15–20
Specialty chemicals	30–35

such as methane and hydrogen. Table 4.4 presents the ranges of energy consumption of each source as percentage of the total use.

Please see the following notes to Table 4.4:

1. Energy sources labeled as Others include import steam, steam transfers, and renewables.
2. By-products are derived from the respective feedstock and consumed onsite.
3. The number for coal is significantly higher in China and other parts of Asia.
4. Consumption of natural gas is for use as energy source only, not as feedstock to any of the processes.

So far, we have learned that the bigger opportunities for energy savings are in the backend of the product value chains and any savings will most likely reduce the use of natural gas as fuel for heating and steam generation. Using electricity as an example, we can analyze in more detail what each form of energy is used for. Table 4.5 shows the use of electricity in petrochemical plants as percentage of the total use of electricity.

The numbers in Table 4.5 show that more than half of the electricity is used for electric drivers and motors, followed by consumption in electro-chemical processes (electrolysis cells, for example), cooling and refrigeration applications, and heating. Other consumption in power generation, process units, and non-process areas is minor. However, later in this section we will elaborate in more detail why even in these areas opportunities for energy consumption can be explored.

TABLE 4.4
Share of Energy Sources as % of Total Energy Use

Energy Source	Energy Use (%)
Electricity	10–20
Fuel oil	1–5
Natural gas	40–50
LPG/NGL	1–2
Coal/Coke	5–10
By-products	10–15
Others	15–20

TABLE 4.5
Use of Electricity as % of Total Use in Petrochemical Plants

User Groups	Users	% Usage
Power generation	Boilers/CHP	1–2
Process users	Heating	5–10
	Cooling/Refrigeration	10–15
	Drivers/Motors	50–60
	Electro-chemical use	15–20
Non-process users	HVAC	1–5
	Lighting	1–5
	Others	1–2

And as a last example of data that will be helpful in analyzing energy efficiency in integrated refining and petrochemical complexes, we can lump all energy consumed in a process scheme for a certain product together and divide it by the production rate of the respective product. This number is referred to as the specific energy use presented in kJ of energy per kg of product. Table 4.6 presents the ranking of the main petrochemical products by their respective specific energy consumption.

The ranking as shown in Table 4.6 shows acetone as the product with the highest specific energy demand (rank #1), followed by polyamine and styrene. Polypropylene, cumene, and acrylonitrile are products with the lowest specific energy

TABLE 4.6
Specific Energy Use by Petrochemical Product

Ranking	Products	Specific Use kJ/kg Product
#1	Acetone	17,500–18,600
#2	Polyamine	9300–10,500
#3	Styrene	8100–9300
#4	Propylene oxide	5800–6400
	Acetic acid	
#5	Terephthalic acid	4750–5800
	Polystyrene	
#6	Methyl tert-butyl ether	3500–4750
	Ethylene oxide	
#7	Polyethylene	2300–3500
	Benzene/Toluene/Xylene	
	Polyvinyl chloride	
	Ethyl benzene	
#8	Polypropylene	<2300
	Cumene	
	Acrylonitrile	

demand (rank #8). Data points like these will be helpful in prioritizing any energy efficiency improvement efforts in a petrochemical or integrated facility.

4.2.1 General Recommendations

Before we dive into the discussion of energy efficiency gains as a result of the integration between refining and petrochemicals, we must mention that there is a huge amount of information on general energy savings programs that can be applied to refining and petrochemicals as standalone efforts and that typically will result in overall energy savings of 5%–10% of the baseline energy consumption. These programs are very detailed and include measures that range from very simple, low-cost changes (for example, a change in an operating procedure) to very complex measures that will require the capital expenditures (for example, installation of a heat exchanger network for heat recovery/integration). In most cases, these capital expenditures show a good return on investment and will pay for themselves within 2–3 years. Below are a few examples of general recommendations for an energy savings program:

- Review and update instrumentation and controls on all fired heaters to optimize parameters such as excess air in the flue gas or draft in the stack.
- Installation of variable speed drives on main pumps and compressors.
- Optimization of reflux rates and reboiler temperatures on distillation columns.
- Implementation of a leak detection and elimination program.

Some of the opportunities discussed in the following integration-specific sections will be also applicable for a standalone facility as certain measures apply to integration between process units in general, not only to a refinery and petrochemicals integration. However, an integrated complex offers a lot more opportunities than a standalone facility due the wider range of processes and available energy sources and users.

4.2.2 Opportunities from Integration

There is a wide range of opportunities to increase the efficiency of energy use and to achieve energy savings stemming from the integration of refining and petrochemicals into one big complex. These opportunities can be organized as follows:

- Energy management
- Monitoring and control systems
- Economy of scale
- Process integration
    - Heat recovery.
- Fuel management
    - Hydrogen management.
- Water management
- Steam management
- Plant air/Instrument air management.

As mentioned before, some of the items mentioned in the following sections may also apply to a standalone refinery or petrochemical facility but may become more attractive or have bigger impact in the integrated complex.

4.2.2.1 Energy Management

The energy management for an integrated complex as well as for any industrial facility has seven main elements that need to be put in place and executed with discipline and rigor in order to achieve the best results from the program. These elements are

- Commitment: everybody in the organization that is running the integrated complex, from the top of the management to the technician in the process unit, must be fully committed to the program, its goals, and its execution.
- Assessment: the detailed study of the performance of each unit and the complex overall is the basis for all goals that need to be set and for all decision-making.
- Action: once the performance assessment has been completed and the goals for the program are set, these goals must be converted to a detailed action plan.
- Implementation: an action plan is only worth something when it is followed, and all proposed actions are implemented in the field.
- Evaluation: following the implementation phase, the performance of the units and the complex overall needs to be evaluated and new results must be compared to the baseline developed in the assessment phase and the goals set in the process (internal benchmarking).
- If the performance is not as desired, the particular action needs to be re-assessed and new, better goals and actions must be developed before these items enter the action phase again.
- If goals are achieved, it is good practice to recognize these achievements. However, this is not a one-way process. Even after achieving a goal, the item or action needs to be re-entered into the continuous improvement process as shown in Figure 4.2.

The visualization of the strategic energy management program enhances the circular character of the main process that allows organizations to continuously re-assess its goals and strategies and to avoid the pitfall of complacency. Doing something right once is a great achievement. But technologies develop, tools get more accurate or more powerful, overall strategies change, and new ideas come up and become feasible. All these developments and others make it necessary for organizations to frequently revisit their assessment of the performance of their facility and set new goals for the program.

The energy management program can greatly benefit from the large amounts of data that are being created, but in most organizations end up unused or only partially analyzed. The latest developments such as cheap computational power, advanced analytics tools, more people with some degree of education in data analytics, and improved data visualization platforms, mining, and analysis of large data volumes become feasible and can be integrated into the energy management strategy as an

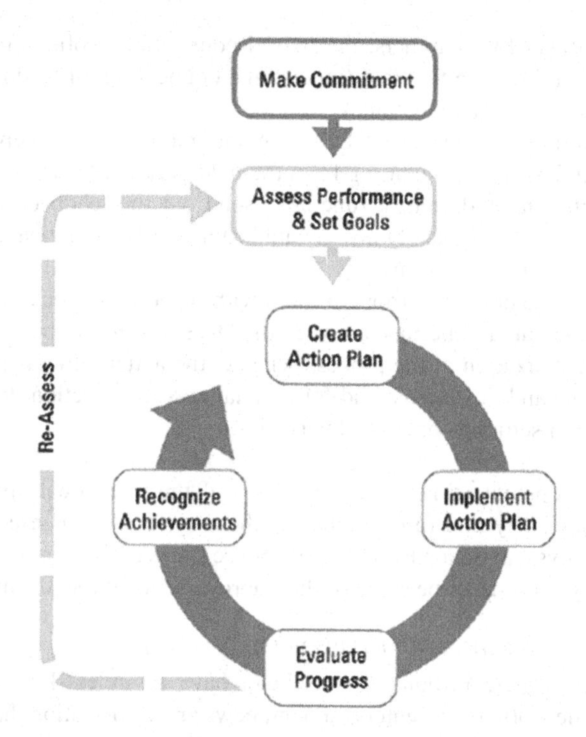

FIGURE 4.2 Main elements of a strategic energy management program.

important step to assess the performance of the facility and set better, more focused goals. Data analytics will raise the understanding of the processes and their interactions. More sophisticated software tools will allow to solve more complex problems or even identify problems that weren't detected using standard assessment methodologies. These could be unknown bottlenecks in the manufacturing process or unprofitable product lines. The implementation of these improved methodologies and technologies will be critical in becoming and staying a benchmark operation, in terms of output, profitability, and energy efficiency.

4.2.2.2 Monitoring and Control Systems

The successful implementation of an energy management program strongly relies on the quality and functionality of the monitoring and control systems in place. These are the systems that read and collect data from the field, compare some data points to setpoints, and send back feedback to control valves and other control elements to apply corrective actions when an actual value diverts from its setpoint. These control systems have several levels that can be described as follows:

- Field instrumentation and control elements: sensors, local indicators, and transmitters; control valves, motor controllers, and other control elements
- Programmable Logic Controllers (PLCs): local control systems that read data from instruments and give direct feedback to control elements

- Input/Output (I/O) cabinets: hardwired connections collecting all information coming from the field (inputs) and connecting all feedback signals going back to the field (outputs)
- Distributed Control System (DCS): computer modules that receive all information and data points coming from the field via the I/O cabinets, perform all calculations and control functions, send out the feedback to the control elements, and typically host a data historian and analytical software to detect any control problems
- DCS interface or control room: a room with several computer consoles that provide the human interface between the DCS and the control room operator for visualization of the process scheme, the actual process parameters, any alarms and deviations, and for manual corrective action, for example, changing of setpoints or control variables.

The rapid development of new technologies and their practical application have changed the flexibility and functionality of these systems immensely and allow a much better analysis of data with shorter response times to the control elements. The following paragraphs describe some of the improvements of the recent years.

4.2.2.2.1 Wireless Sensors and Transmitters

In these modern times, millions of households have a wireless Internet connection and access to the world of the entertainment, news and information that is the World Wide Web. So, it is no surprise that wireless technology has found its way into manufacturing facilities such as refineries and petrochemical plants. Wireless transmitters started replacing the hardwired versions, allowing the operations and maintenance teams to take advantage of the increased accessibility and flexibility of wireless signals. If a transmitter is hardwired via the I/O room to the control system, only the control system can read the signal and interpret it. Other access to the data will have to go through the control system itself. With a wireless signal, multiple people can access the same data that is transmitted to the control system, either via a personal computer, an intrinsically safe tablet or cell phone. This allows for instant review of data points, more flexible use of analytical tools, and for reduction in response times to data deviations from setpoints or allowable data ranges.

4.2.2.2.2 Industrial Internet of Things (IIoT)/Digital Twin

Wireless sensors also allow for an automated and instant upload of large data quantities onto cloud-based data management systems, where the data can be stored, accessed remotely, and analyzed using the latest tools and software, without impacting the capacity and performance of the actual control system. The connection of identified electronical components and computers and calculation devices is also referred to as the Internet of Things, or in this case the Industrial Internet of Things. It describes a world in which electronic devices communicate with each other without any human interaction. More and more manufacturing companies move toward an actual digital representation of their process or facility, the so-called digital twin. A digital twin will allow better visualization of complex issues and – if connected to a dynamic simulator – the testing of proposed changes before they are entered in the control system.

As you can imagine, these developments also raised a whole new level of concern around cyber-attacks, hackers and industrial espionage. These issues need to be addressed carefully to secure all proprietary data and intellectual property from outside attacks.

4.2.2.2.3 Enhanced Advance Controls

Advanced controls were a big thing in the 1990s with the first wave of technologies that were intended to support the human operator in making better decisions faster. The technological advancements of recent years and the introduction of what is now classified as artificial intelligence (AI) have many companies revisit their advanced control strategies and the technologies currently applied. To make a distinction between the first wave of advanced control applications and the new technologies, we call it enhanced advanced control. The principle is still the same as new application still uses a neural network approach and knowledge-based fuzzy logic as the base building block for the enhanced advanced control application. In simple words, these systems mimic the reactions and responses of the best in class operator, just in a time, with a frequency, and with an accuracy that is beyond human performance. The performance of the best in class operator must be collected by interviews and simulations and then translated into the digital world. Once this step and other preparatory testing have been completed and the system is online, its machine learning capabilities will allow it to learn from experience, improve its responses, and become even better than the best in class operator.

In reference cases presented at multiple conferences and technical meetings, companies reported energy savings ranging from 2% to 18% with the implementation of an improved energy monitoring and control system. These savings come from improvements in balancing energy sources to minimize consumption of primary energy sources while satisfying the energy demand from all process systems. This can apply to the intelligent balancing of the electrical power grids, balancing of fuels, balancing of steam generation for the different pressure levels of steam, and the monitoring and minimization of energy demand from the end users.

As with every automated system, enhanced advanced control is only as good as its input and the data it receives. Frequent failures by the system to initiate the correct response will destroy the confidence of the control room operator(s) in the system. An operator who is not confident in the capability of the system will tend to switch it off and manually control the process. This will eliminate any benefit that can be gained from the implementation of enhanced advanced control. Therefore, be very detailed, very diligent and very careful in the development of the basis of the system as even the smallest mistakes or oversights will cause the system to fail and rob it from the opportunity to learn.

4.2.2.3 Economy of Scale

The term "economy of scale" comes from capital projects and describes the fact that the specific cost to design and install an industrial facility (or other facility, for that matter) will go down the bigger of a facility you build. In other words, the amount of money you spend per unit of plant capacity decreases with the size of the facility. This principle also applies to utilities and power generation. For example, combining the demand for electricity for the refinery with the demand for the petrochemical plants will allow the construction and operation of a bigger power plant. Some of the cost to construct such a plant is fixed or doesn't vary much with the size of the plant, giving the bigger plant an advantage from the start. In many cases, bigger systems can be operated more efficiently. This is especially the case when the demand for a certain energy such as steam or electricity varies. By finding smart ways of using load factors, multiple parallel trains, and smart systems for load balancing, the bigger system will allow the realization of energy savings over the smaller system. Another approach to lower energy cost in a bigger facility is combined heat and power (CHP) generation. This concept takes advantage of the efficiency gain by combining the generation of heat for the process units with the generation of power. In small scale, this concept often doesn't make economic sense due to the cost of the equipment and the complexity of the generation process. In a large-scale application, CHP becomes more attractive and may be utilized to lower the cost of energy even more. In general, by using common utilities for the large, integrated complex, the variable specific cost for energy generation will be lower. And this is one of the pieces of the increase in competitiveness on the fuels and petrochemicals markets.

4.2.2.4 Process Integration

We already talked a lot about process integration when we talked about the different feed and product streams. Some products of the refinery will be feed streams to the petrochemical units. And some by-products of the petrochemical units will be feed streams to refinery units. The direct coupling of process streams will allow the integrated complex to take advantage of the following effects.

4.2.2.4.1 Potential Reduction in Intermediate Product Storage Volume Requirements

This will reduce the cost of operation. Feeding intermediate products directly into the downstream process units will reduce the need for storage space, for loading/offloading, or for pipeline transport.

4.2.2.4.2 Elimination of Pumping Services

Process integration will allow for the elimination of storage and product loading-/offloading-related pumping services, or at least reduce the use of certain pumps significantly. With the reduced storage space requirements, intermediate pumping services may be eliminated, or at least the use of these services will be significantly reduced in frequency.

4.2.2.4.3 Reduction of the Need for Product Cooling to Storage and Reheating from Storage

Hot feeding of intermediate products to downstream processing units means that the product will leave the delivering unit at high temperature and bypass the heat exchangers and air coolers that were necessary to bring the product down to storage temperature. Then, it will bypass the heat exchangers in the feed preheat system of the receiving unit and, in some cases, significantly reduce the required energy input to fired feed heaters.

4.2.2.4.4 Heat Integration/Heat Recovery

An integrated complex offers more opportunities to use a hot stream from one unit to heat up another stream in another unit. The same applies to using a cold stream of one unit to cool down another stream from another unit. The most common tool to develop the optimum concept for heat integration and heat recovery is the pinch analysis. Pinch analysis is based on the development of composite curves for heat sources and heat sinks. These curves can be created for specific units or for the complete complex. An example is shown in Figure 4.3.

The upper composite curve represents the available heat sources in kW at their respective temperature. The lower composite curve represents the heat sinks in kW at their respective temperature. The area between the two composite curves that is shaded in gray is the amount of heat that can be recovered and transferred between the hot and the cold streams. The delta between the end points of the two curves on the left, lower end is the amount of cooling that needs to be provided externally. External cooling means the use of cooling water and air. The delta between the end points of the curves right, upper end represents the amount of heating required. Heating can be provided by steam generation or direct fired heaters.

The difference between the two curves is the temperature difference and driving force for the heat exchange. It is common for the design of heat exchangers of any type to define a minimum temperature difference to make sure that sufficient driving force will be available to keep the size of heat exchangers, more specifically the area for heat exchange, in a range that is feasible for fabrication and justification. The points where the lines are closest to each other and may approach or even go below that minimum temperature difference are called pinch points.

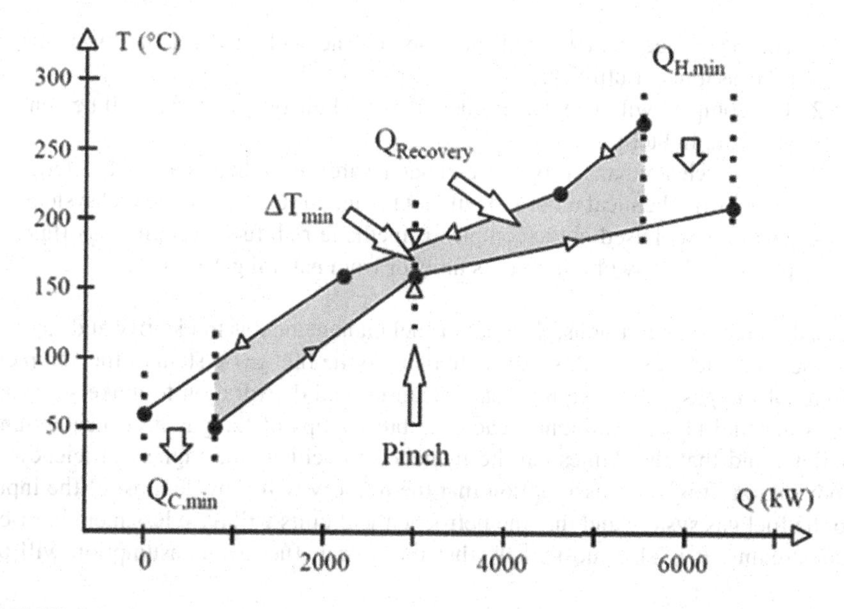

FIGURE 4.3 Example composite curves for pinch analysis.

Eliminating these pinch points by rearranging the heat exchanger configuration or the sequence of heat exchangers for heating/cooling will allow for an increase in recoverable heat and, consequently, a reduction in heating and cooling requirements.

With the increased computing power and improved software tools, pinch analysis can be easily done to great detail and heat integration can be performed to an extreme where it becomes impossible to implement in the practical world. Installation of numerous small heat exchangers to recover every kW of energy theoretically available bears the risk of spending capital at the wrong end. Also, extreme heat integration between units can be an operational nightmare as you might encounter a case where a process unit can't be started up or operated without one or more other process units being in operation and providing the heat needed for integration. If you have several units that are integrated to an extreme degree, the unit with the lowest level of reliability and the highest rate of rate reductions will drag down all other units as they will have to reduce rates or even shut down with the bad acting unit. And providing a back-up system or an alternative system just defeats the purpose of the integration. While heat integration and heat recovery are great opportunities to increase the energy efficiency of a facility, going too far with the integration can create operational and reliability problems that will negate the positive effects of the integrated concept.

And as we had mentioned before, pinch technology has been developed to a point where its principle can be applied not only to heat and energy, but also to molecules such as water and hydrogen.

4.2.2.5 Fuel Management

Before we discuss the actual management of the fuel system, let us agree on a few assumptions that will simplify the fuel system significantly.

1. The integrated complex will not use coal nor coke as fuel in any of the process units or utilities.
2. The complex will also use no fuel oil as fuel oil-type material will be converted to lighter products.
3. Gases such as ethane, propane, and butane are valued higher as feed material to the petrochemical units and will not be sent to supplement the fuel system.
4. The only fuel used in the complex is methane-rich fuel gas either as offgas produced by any of the process units or from natural gas import.

Based on these assumptions, the goal of fuel management is to identify and capture all methane-rich gas streams and route them to the fuel gas system of the complex. Some of the gases may require some clean-up and dehydration to make sure that no water and no contaminants reach the burner tips of the gas-fired heaters and boilers, and that the flames can be managed to achieve the highest efficiency of combustion. It is a fair assumption that the refinery will provide most of the input to the fuel gas system and that the petrochemical units will have less methane-rich gas streams. It is also most likely that the overall fuel gas consumption will be higher than the amount of internally produced methane-rich gas. In some cases, methane might be valued higher as a feed stream to the petrochemicals side, for example, for production of methanol. The fuel management system, which is a sub-system of the process control system of the facility, will detect the demand for fuel gas and balance it against all available methane-rich gases and provide the gap in the balance by importing natural gas. An automated system will allow the complex to minimize the import of natural gas and maximize the use of available internal fuel gas streams. As mentioned earlier, the fuel management system will also monitor the performance of each fuel gas user and ensure that all parameters are kept at their optimum operating point to maximize the efficiency of fuel gas usage throughout the facility.

4.2.2.5.1 Hydrogen Management

When it comes to hydrogen, it becomes clear that a refinery and a petrochemical facility are a match made in heaven. Refineries are permanently short on hydrogen as they must meet very stringent sulfur specifications in their liquid fuels which are adjusted to lower values on a regular basis to meet the aggressive goals for lowering emissions from combustion of fossil fuels. Therefore, refineries have to either find sources for import of hydrogen or spend a lot of capital for and carry the burden of operational expenses for hydrogen manufacturing via SMR as well as hydrogen purification via PSA or membrane separation.

On the other hand, petrochemical facilities have excess hydrogen from processes such as PDH and typically export the hydrogen into a pipeline.

The hydrogen management system for a combined complex will allow the operations team to

- Maximize the recovery of all hydrogen-rich streams from the refinery and the petrochemicals facility
- Monitor the hydrogen purification section
- Balance hydrogen demand for desulfurization, hydrogenation, and cracking units
- Minimize or eliminate the need for import of high-purity make-up hydrogen.

To achieve the maximum flexibility from the internal hydrogen network, the complex will have to add piping, controls, and compression power. This represents additional capital expenses that will easily be compensated by the savings in reduced hydrogen manufacturing and purification cost.

4.2.2.6 Water Management

Why do we talk about water management in the context of energy efficiency? First, water is used in the energy system of the integrated complex as feed to the steam generation and as heat sink via the cooling water circuit. So, water plays a significant role in the energy balance of the complex. Second, in some areas water threatens to become a scarce resource and careful consideration must be given to the water balance, the use of fresh or make-up water for the complex to achieve a sustainable operation.

To make sure that the integrated complex fits into the environment and the overall water balance of the area it is either located or planned to be located in, specialized consultants can be engaged that have tailored tools and models to assess the impact of the integrated facility on the regional water balance. These models take into consideration:

- Neighboring and close-by industrial facilities, their water usage, and wastewater production
- Neighboring and close-by residential areas, villages and cities, and their water usage and wastewater production
- Neighboring and close-by farms and agricultural facilities, and their water demand and wastewater production
- Neighboring and close-by forests, grasslands and parks, and their water demand
- All close-by water sources such as oceans, rivers, lakes, creeks, groundwater reservoirs, and other aquifers
- All close-by wastewater treatment plants and their respective capacities.

These models can be used to simulate the impact of the additional water demand from the integrated complex on the level and sustainability of the available water sources, to assess the impact on other water users to ensure no other users such as small communities or farms get starved from their water needs, to ensure that the existing wastewater treatment capacity is sufficient to handle the additional load from the integrated facility or identify capacity gaps that need to be filled, and to assess options for water recycling and reuse in the integrated complex.

Soon, industrial complexes such as an integrated refinery/petrochemicals facility will start following the concepts applied to smart city designs and look at opportunities to recycle and reuse more water up to the point of reaching a closed-loop and zero-discharge water system.

One key component of good water management is the separation and targeted treatment of different wastewater streams which will allow better results in water treatment and resulting water quality for reuse. If wastewater streams carrying certain contaminants can be separated, these contaminants can be removed from the water and recycled into the process. Examples are

- Potassium hydroxide and its use in hydrofluoric acid (HF) alkylation units
- Biodegradable polyesters and their use in the manufacturing of polyhydroxyalkanoates
- Ammonia and its use in ammonium sulfate fertilizer.

The biggest improvements are expected from developments in membrane-based water treatment, but other treatment steps will improve in effectiveness as well.

By-products from water treatment, mostly sludges, can be utilized in the generation of heat for the facility to further reduce the amount of waste to be discharged from the complex.

4.2.2.7 Steam Management

The typical steam generation system in an integrated complex will have a water treatment section to remove hardness and salts from the water and produce demineralized water. These sections are often called water softeners or water polishers. The demineralized water is then preheated and fed into a steam drum, from where it circulates through a steam generator and vaporizes. A simplified flow scheme is shown in Figure 4.4.

A steam boiler can be indirectly fired with fuel gas or use recovered heat from a hot process steam. Due to the different sources of heat and the temperature at which they can generate steam, the complex will have at a minimum three standard steam levels we see in refineries and petrochemical plants:

- Low-pressure steam at 3.5 barg (<3.5 barg)
- Medium-pressure steam at 15 barg (3.5–17.5 barg)
- High-pressure steam at 30 barg (>17.5 barg).

These pressure levels may vary as indicated by the ranges in brackets. The optimization of the number and pressure of steam levels coming from the pinch analysis may result in a higher number of steam systems that will allow to supply heat more targeted to specific users. However, if the number of steam levels gets too high, it will complicate the operation of the facility as well as increase the cost for maintenance

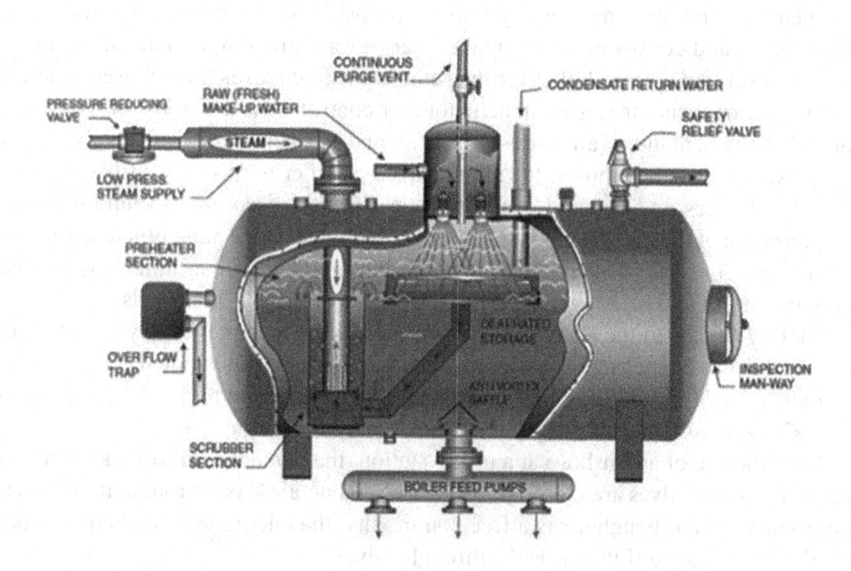

FIGURE 4.4 Simplified steam generation schematic.

and spare parts as the design pressures will require different parts that won't be easily or not interchangeable between systems.

Should the steam balance between the systems require letdown stations, the letdown can be accomplished by small steam turbines to generate electricity instead of simple letdown valves. The cost of procurement, installation, and maintenance must be evaluated against the reduction of fuel use for power generation.

Combination of the steam demand from the refinery and the petrochemical plants will allow for installation of a bigger, more efficient steam generation system that will reduce the specific fuel demand per unit of steam generated.

The balancing of more steam users will also reduce the overall impact of fluctuations at one unit on the overall demand which leads to a more stable workload, resulting in a more efficient operation.

Other measures that can be taken to improve the efficiency of the steam system include

- Elimination of stand-by losses and steam venting to provide excess capacity as back-up
- Improved condensate return rate to reduce the amount of make-up water needed
    - Elimination of leaks in the water system and in the steam systems, especially at and around steam traps.

The implementation of a CHP system will increase the utilization of internal streams carrying waste heat for the generation of power and steam.

4.2.2.8 Plant Air/Instrument Air Management

The plant air and instrument air system are probably the simplest utility systems in the complex and consist of an air intake filter, several air compressors, an air dryer, and a pressurized storage bullet for instrument air. Plant air is mostly used at utility stations. Instrument air is used in actuators for control valves. The list of users indicates that the demand of air varies over time and frequently as control valves open and close or as air is required for maintenance or other purposes at utility stations. A study of users and demand cycles will determine the optimum solution for the configuration of the air system. Bigger compressors will work more efficiently if providing a constant base load to the air system, but multiple smaller units may provide more flexibility and higher efficiency at lower, more variable workloads.

Cooling the air to the right temperature will increase the efficiency of the moisture removal as condensate is easier to separate from air than vapor. This will allow to control the dew point of the instrument air to a point where reliability of actuators is maximized, and moisture-related malfunctions can be eliminated.

After the use of an air hose at a utility station, the personnel must make sure that the air isolation valves are closed properly and that no air leaks through the valves to atmosphere. Even though air is a free commodity, the energy used to compress and dry the air is wasted if the air leaks through valves.

4.2.3 Human Factor

When it comes to day-to-day operation and the monitoring and use of the automated control and management systems we described in the sections above, we still rely on operators and technicians, therefore, on human beings to make the right decisions and use the systems for the purpose they are provided for: the execution of repetitive tasks that will free up time for the human workforce to focus on other tasks. Some of these tasks are critical to the success of the overall energy management program, for example:

- Only operate energy users when you need them. Don't switch them on too early, and don't let them run long after you stopped needing them. This applies to pumps, compressors, fans, lights, air conditioning, heating, cooling, and any other energy user you may operate at your place of work.
- Report any leaks you observe, may it be water, oil, gas, steam, or air. Make sure the leak is repaired as quickly as possible.
- Control your process variables to the value they need to be at for optimized operation. Don't manipulate pressures, temperatures, or levels to a point where operation is convenient for you but far off the optimum operating point. This rule may also apply to the temperature setting of the air conditioner in your office.
- Religiously execute the required predictive and preventative maintenance programs to keep equipment and components in a condition that allows them to operate safely, reliably, and efficiently.

This is the reason why the first step of implementing an energy management program is the commitment of all employees on all levels, in all departments, of all disciplines to actively support and execute whatever is identified as an actionable item.

4.3 TECHNOLOGY DEVELOPMENT

Both refining and petrochemical facilities rely on solid and mature technologies for the production process, which are continuously optimized to keep up with competitive scale economies. Latest processes have implemented several improvements to minimize the energy consumption and capital cost; however, another optimization path goes by creating integrated schemes, but unfortunately, such integration cannot be standardized because the options for integration are different case to case as discussed in Section 4.2.

This section will summarize potential routes and available technologies and the challenges to achieve such optimization with emphasis in their flexibility to work in an integrated scheme.

4.3.1 Flow Schemes for Integration

As described in Section 4.1, the integration schemes will vary from different locations, and even for specific regions, it will depend on the starting point of each facility (new petrochemical facilities plugged into a refinery, new refining units plugged into a petrochemical complex, existing integrated systems, or a full grassroots configuration). An overall integrated flowsheet is presented in Figure 4.5, to illustrate all possible options and the corresponding interconnections between them.

4.3.1.1 Refining Products Flowing to Petrochemical Units

LPG. This is a valuable stream to produce petrochemicals, and depending on the region, it will flow as direct feed to the crackers or as feedstock for the generation of propylene; however, when alternative routes are available for olefins production, the stream is still energetically valuable as fuel.

Naphtha/Gas Oil. Similarly to LPG, these are valuable feedstocks for a cracker, but in the case of the naphtha, the value can also support the production of aromatics, and different factors may affect the decision, i.e., availability of lighter streams for cracking (ethane), volume of aromatics demand in comparison with fuel demand, etc.

C4's. This is a very flexible stream to fill gaps on the schemes once other priority processes have been defined; however, where a strong demand for SBR (Styrene Butadiene Rubber) calls for the maximization of butadiene production, this stream will be treated with special interest. Regardless of this decision, other C4 components may have other potential destinations (i.e., butenes to produce propylene by metathesis)

Reformate and Aromatics. Giving the flexibility of catalytic reformers and catalytic cracking units, this stream may flow toward the aromatic extraction unit or the gasoline pool, and in this case, the seasonal demand of fuels may have the greater influence on the operational mode.

4.3.1.2 Petrochemical Products Flowing to Refining Units

Hydrogen. As previously discussed in Chapter 2, hydrogen is one of the main streams bridging between petrochemical and refining units especially for refineries in locations with high environmental restrictions where a considerable level of hydrotreating is required. For integrations in existing facilities where enough onsite production is available, the value will be quantified as fuel to be exported or depending on the volume and location of nearby facilities as feedstock (i.e., ammonia production).

Pygas. This is another flexible stream, and its final destination will depend on the desired level of integration. When an aromatic complex is part of the scheme, it is very likely to keep this stream as petrochemical feedstock, but if aromatics production is not in the game or is not a priority, the preference may be inclined for its blending into the gasoline pool.

Heavy Aromatics. Toluene or similar aromatic components provide a considerable value in the gasoline pool from their high octane rating; in facilities where its processing as aromatics may compete with the fuel demand, the decision may come from the cost of processing these streams and the value of final products (i.e., transalkylation to increase the volume of paraxylene).

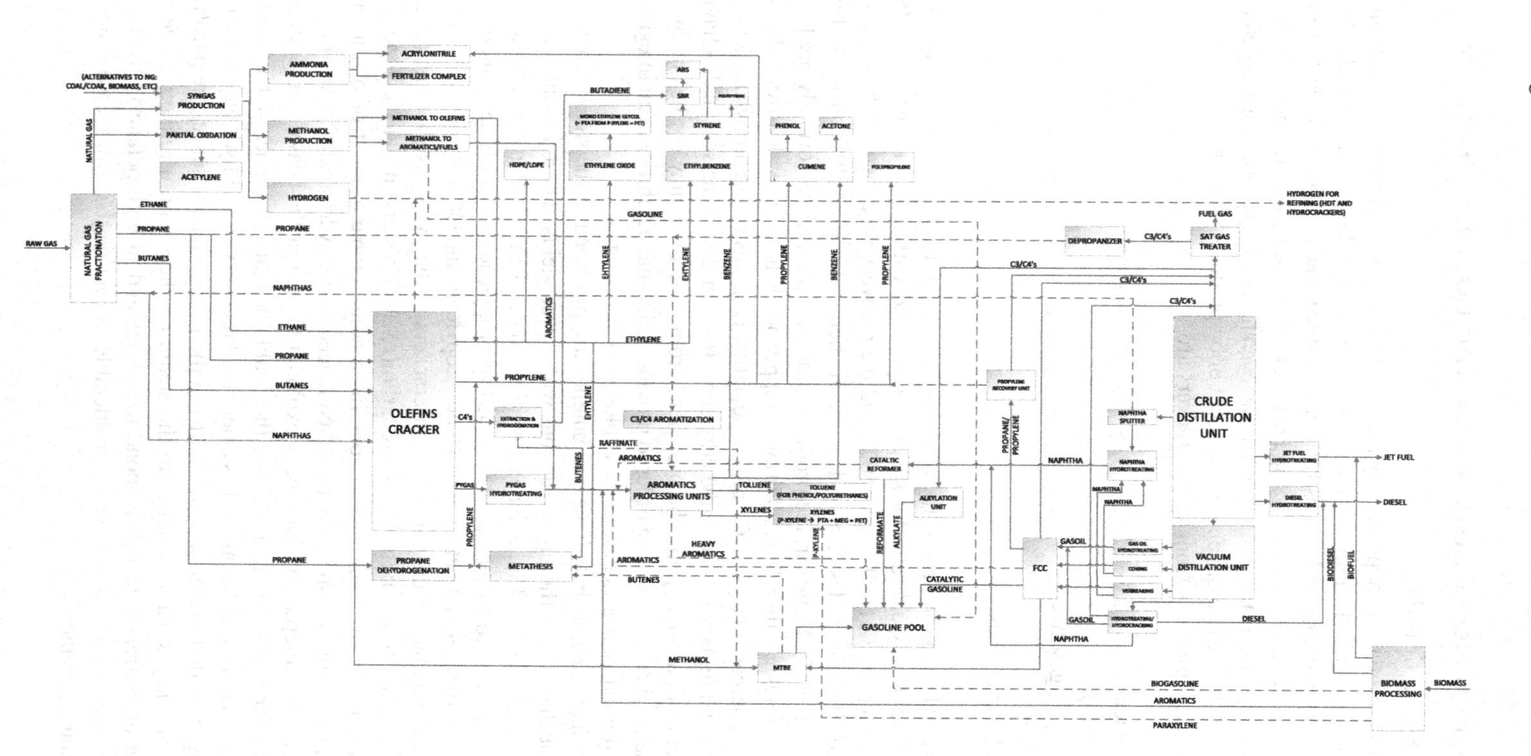

FIGURE 4.5 Schemes for Integration.

4.3.2 Role of Steam Crackers and Catalytic Crackers on Integration

4.3.2.1 Steam Cracking

Steam crackers are the core processing facility of the petrochemical industry, and their value comes from the flexibility to handle different feedstocks with few modifications in the process configuration but mostly because of the broad number of products resulting from this process. The heavier the feedstock, the wider the range of by-products that can be recovered in the steam cracker. But even for crackers processing only ethane (the lightest feedstock possible), some amount of by-products is generated.

The main by-products from steam crackers are

- Propylene
- C4 stream
- Pyrolysis gasoline
- Hydrogen
- Acetylenic gases.

After the ethylene, propylene is the most important product from the steam cracker. Propylene/ethylene yield ratio for steam crackers fed with heavy feedstocks (LPG, naphtha, etc.) is between 40% and 50%, and the recovery sequence can be designed to produce different grades of propylene either chemical grade (~90% purity) or high purity (above 99%). However, the production of ethylene will be normally predominant over propylene; therefore, additional supply needs to be provided from alternative supply sources such as FCC units, especially in regions where propylene demand is growing at a higher pace than the ethylene. However due to the seasonal behavior of fuel demand, at some point the propylene from FCC units may not be enough to supply the market; therefore, in order to provide additional flexibility in an integrated scheme, alternative routes have to be explored to fill this gap.

C4 products such as butadiene and butenes are produced in the steam cracker in similar mechanism as the propylene, but with lower yields. These products have a specific market (i.e., 1–3 butadiene); however, some of the C4 components such as 2-butene can be used to increase the yields of propylene, and it all depends on the desired yield distribution between ethylene, propylene, and C4's.

Another valuable product from steam crackers is the pyrolysis gasoline (pygas) which is essentially a mixture of hydrocarbons with C5's and heavier components rich in aromatics; however, it also contains a considerable amount of olefins and diolefins; therefore, at some point this stream needs to be hydrotreated to saturate the olefins. Possible alternatives have been explored to recover the heavy olefins stream to produce propylene, which introduce the double benefit of a higher global yield of propylene from the cracker as well as a better net recovery rate of hydrogen since this is no longer needed to saturate the olefins in the pygas stream. Hence, the pyrolysis gasoline is a key stream to integrate the steam crackers with a refinery since the definition of its treatment will respond to the demand in high octane vs. rich aromatic streams but will also have a considerable impact on the hydrogen balance in an integrated scheme.

The cracking of saturated alkenes results on the release of hydrogen which can be recovered with a purity in the order of 95% but with a low level of contaminants. Therefore, it is a valuable feedstock for refining units, and its contribution is directly beneficial to the demand of onsite generation through steam reforming units which are a high consumer of energy in the refineries.

Acetylenic components are normally an undesired co-product resulting from over-cracking of the feedstock and, therefore, are normally hydrogenated and recycled to increase the yields of the corresponding olefins; during this process, heavier olefines such as butadiene can also be saturated; therefore, if the acetylenic components are not desired, the recovery of C4 streams is needed upstream of hydrogenation reactors. However, if the cracker is built in a region with considerable demand of these gases, the process can be tailored to achieve specific yields of the products. These components are particularly hazardous due to their high reactivity; therefore, sometimes it's preferred to leave these components as part of heavy streams from the cracker to keep their concentrations diluted to the lowest level possible.

4.3.2.1.1 Limitations of Steam Crackers

The pyrolysis of hydrocarbons is an endothermic process, and highly favored by high temperatures, which lead to the formation of a substantial amount of coke that needs to be removed frequently; therefore, steam crackers are normally designed with multiple pyrolysis furnaces to allow for coke cleaning. Main consequences are the economic impact for downtime and the increase capital cost to provide spare furnaces.

Another downside of high-temperature operation in the reactors is the consequent impact on energy demand. Typical energy consumption in an ethane cracker is in the order of 15 GJ/MT of ethylene where at least 60% of this amount is consumed in the pyrolysis furnaces. Most of this energy is generated burning fuel gases produced in the same reactor as by-product. Some of the energy can also be recovered producing steam which can be further used to run compressors in the compression and refrigeration systems; however, a potential risk is the high amount of CO_2 released to the atmosphere during the combustion in the furnaces, and stringent regulations in the future may push the olefins producers to look for improved processes to reduce these emissions. Some of these technologies not only will benefit the environmental position of olefins production facilities but also may contribute to a better integration with other technologies and to increased yields of propylene and aromatics.

Keeping in mind that the main driving force for energy consumption in the steam crackers is the need for high temperatures in the reactors, the logical solution would be to reduce this requirement which can be achieved by the use of efficient catalytic process; however, conventional catalytic cracking would narrow the profile of by-products from the reactor effluent reducing considerable profit and flexibility in the process. Different catalytic processes to replace or improve steam cracking have been explored but still not mature enough to provide the benefits of steam cracking at a lower energetic and environmental price, and these will be described in the following section.

4.3.2.2 Catalytic Cracking

Most of the catalytic cracking processes available either commercially or in development take as a reference the model of the FCC process, and most of the work has been focused on the improvement of the catalyst of existing FCC configurations and in some cases also redesigning the configuration of reactors. But in general, all catalytic processes work at lower temperatures, and without (or minimum) downtimes for de-coking. Similarly to the FCC process, these alternatives are suitable for regions with availability of heavy feedstocks, but the challenge is to fall in the feedstock range that is not competing with steam cracking (i.e., using naphtha as feedstock).

Any new catalytic process needs to be designed with total flexibility to receive a wide range of feedstocks to be capable to adjust to market demand unbalances between ethylene, propylene, and fuels. One successful example is the K-COT technology developed by KBR, which evolved from their initial FCC reactor design (Orthoflex) implementing different improvements to the catalyst (i.e., the SK catalyst implemented in the Advanced Catalytic Olefins process) to increase up to 25% higher yields to olefins compared to the steam cracking process but also increasing the propylene to ethylene ratio between 1:1 which in a typical steam cracker is restricted in the order of 0.6:1 as maximum. Another advantage is the increased production of aromatics and the great flexibility to receive feedstocks from different sources; the benefit of operating at lower temperatures than in steam cracking is reflected in a reduction of specific energy consumption in the order of 10%, and the purification system is similar to the steam crackers, but heavy streams can be recirculated to achieve maximum conversion.

Another path of development has led to the improvements in the configuration of reactors, one example is the replacement of risers with down flow reactors, and this is the result of a combined effort between Saudi Aramco, JX Nippon Oil & Energy Corp.(JX), King Fahd University of Petroleum and Minerals, Technip, and Axens. The use of down flow configuration helps to reduce back-mixing between catalyst and feed which allows the reactor to handle higher catalyst/oil (C/O) ratios and therefore is possible to increase the reaction temperature that not only favors the conversion of olefins with the selected catalyst but also allows for shorter residence times to reduce thermal cracking. This technology bumps the olefins conversion up to four times compared to the conventional FCC process and can be operated in high octane or olefins mode. In both cases, the olefins yield is higher than for the conventional FCC. This flexibility makes this process a great candidate for integration with refining.

Another improvement in the reactor topology to achieve better yields of propylene is the addition of a second riser; example of this configuration is the MAXOFIN technology from KBR or the PetroRiser in the Axens FCC process.

Another route of interest in catalytic cracking is the processing of heavy feedstocks, and these technologies are mainly attractive in regions with vast number of heavy feedstocks such as China. In that regard, some important contributions have been developed by the Chinese Sinopec Research Institute of Petroleum Processing (RIPP), and one of these is the Deep Cracking Catalytic process licensed by Technip outside China. The process works in similar conditions to the FCC process with temperatures around 500°C–600°C and low pressure, and steam consumption is higher than for the conventional FCC, but considerably lower than the steam demand of the steam cracking process. DCC has enough flexibility to work as olefins production unit (Type I) or also closer to an FCC unit (Type II). When working as Type I propylene yields are in the order of 20% and the process can be integrated with a steam cracker sharing the fractionation and recovery system to minimize operational cost and recirculate the maximum amount of high-value products. Another example is the catalytic pyrolysis process which has been explored extensively in late years; the main advantage is their potential to process not only heavy hydrocarbon fractions but it's been proved to achieve yields of propylene and ethylene in the order of 15% and 10% correspondingly by using biomass as feedstock.

In general, the FCC process can be tailored to get some degree of flexibility between gasoline and olefins production; some examples are the INDMax FCC process from CB&I/Lummus or the Petro FCC from UOP that running at higher temperatures and C/O ratios can deliver yields close to 25% of propylene with a lower carbon footprint compared to a steam cracker; but still the volumes may not be considerable enough to meet the demands that a steam cracker should provide especially for propylene. Therefore, supplemental catalytic processes may be required to close this gap like the interconversion processes discussed in Chapter 2; an example of these technologies is the Propylur® process from Lurgi or the Olefin Cracking Process (OCP) from UOP which uses a fixed bed reactor to produce light olefines from a heavy olefins stream coming from any of the FCC-type process previously described or from Steam Crackers; the effluent from these reactors is rich in light olefins with a high propylene/ethylene ratio maximizing the overall yields of these components in an integrated configuration; this process requires cyclic regeneration, but similar technologies have been developed by ExxonMobil (MOI and PCC) using fluidized beds.

4.3.3 Role of Catalytic Reformers and Steam Crackers on the Integration of Aromatic Streams

We have outlined how different olefins enhancement process supplement the gaps from steam crackers; however, some of these processes may have an impact on the overall yields of aromatics; therefore, it is important to consider the role of other sources of aromatics and their potential interfaces with steam crackers and alternative olefin technologies.

The FCC process is one of the main sources of olefins, but also it represents an important catalytic route for the reforming of naphthas, contributing in the supply of aromatics, which is supplemented by the aromatic production from the steam cracking. This is opposite to the production of the olefins where the steam cracker is the main source supplemented by other catalytic processes.

The reforming process involves the conversion of paraffins into iso-paraffins and naphthenes with further dehydrogenation to produce aromatic components. This is a well-developed and mature technology with several licensors available in the market like UOP with the CCR Platforming, Axens with the CCR Aromizing or a more recent incorporation from the Chinese Sinopec. But in general, all of them follow a similar process sequence based on a set of highly endothermic reactors operating at very low pressures with a considerable amount of hydrogen (yields close to ~5%).

Catalytic reforming units are a key piece for integration of different streams since these can receive feedstock directly from crude distillation units, or indirectly from hydrocracking or FCC (or FCC type) units processing heavy feedstocks such as vacuum gasoil or light cycle oil to produce heavy naphtha to reform into aromatics. This connection is critical and provides a great degree of flexibility allowing a more open integration scheme.

On the other hand, catalytic reforming units are also an important contributor of blend material to the gasoline pool; therefore, there is a competition between aromatic derivatives and high octane gasoline. One of the main differences in the operational modes is the need of removal of C6 components in the feed when gasoline is the preferred product, because this will yield to benzene that is an undesired component in the gasoline, while in aromatics mode, this requirement needs to be switched. Over the time, an increase and sustained demand of aromatics is expected; therefore, a transitional period is expected where the demand of fuels and aromatics will coexist pushing the design of catalytic reformers to provide enough flexibility to handle different operational modes imposed by the seasonal demand of fuels.

The pygas recovered from the steam crackers has a considerable amount of aromatics; however, it also comes with olefinic components and some amount of contaminants (i.e., sulfur and nitrogen). These need to be removed before the aromatics stream is sent to fractionation. These components are removed in a two-stage hydrotreating process.

4.3.4 Impact of New Developments in Technologies

Petrochemical and refining industries are based on very well-developed technologies; however, in order to respond to different limitations and external restrictions (i.e., environmental, commercial, etc.), there is still a continuous research in the field of new alternative processes, some of them more promising than others; but at some point, their incorporation as a commercial technology will have an impact (either negative or positive) on current integration schemes. Some of these potential technologies are summarized in this section.

4.3.4.1 Oxidative Dehydrogenation

The dehydrogenation of olefins has been explored extensively especially for the production of propylene as an option of on-purpose generation of this product. The main driver to explore this option is the fact of having substantial amount of natural gas available (and the corresponding associate liquids) as in the case of the USA where the abundance of shale gas makes ethane and propane attractive feedstocks to justify the use of these components to produce olefins. But the dehydrogenation reaction, similarly to the steam cracking, is an endothermic process and conducted over catalytic conditions; this process is very demanding in terms of energy, especially for ethane where temperatures over 800 °C are required.

However, the oxidative dehydrogenation is on the contrary an exothermic process that can be carried out at lower temperatures, with a considerable reduction to the energy consumption in the order of 30% compared to the energy required in the steam cracking process, and therefore would apply as an ideal candidate to improve the operational cost in the production of olefins. Also as a secondary advantage, the coking is substantially reduced due to the presence of oxygen which promotes its conversion into carbon dioxide, improving the on-stream times.

However, the use of oxygen in the process, has a negative effect in the production of carbon dioxide not only produced from de-coking but also from the reaction itself. Therefore carbon monoxide (instead of oxygen) has been proposed as oxidant introducing a double benefit, the carbon monoxide required in the process can be collected from the capture in another neighbor unit but also since the by-product is now carbon monoxide, it opens a potential integration with a syngas user in another facility (i.e., a hydrogen unit).

This technology is in early stages of development for ethane but available with some maturity in propane; nevertheless, the availability of ethane and propane from shale gas represents a great incentive to further explore this technology.

4.3.4.2 Biomass Processes

As described in Section 4.1, biomass represents an option for decarbonization and a potential alternative to replace fossil fuels; however, the incentives to implement these technologies are in a very early stage and at this point depend on the alignment of several conditions such as more stringent environmental restrictions, very high crude oil prices, and development of other alternative technologies (i.e., methanol to olefins or gasoline).

However, a considerable progress has been made to define the possible routes for its implementation; in general, four lines of development can be identified:

- Gasification

  This route is focused on the production of syngas from biomass which can further be transformed into fuels through the Fischer–Tropsch process, or into methanol or ethanol using a catalytic process, and then, the alcohols can be converted into olefins by dehydration.
- Catalytic Pyrolysis

  This option involves the hydro-deoxygenation of the biomass under catalytic conditions with further hydrotreating, and a successful example of this implementation was developed by the Gas Technology Institute now in collaboration with Shell to produce fuels.
- Hydrogenation and Reforming

  In this process, the target is to deoxygenate the feedstock by hydrogen enrichment with a subsequent reforming process in aqueous phase that produces fuels and aromatic components, and such process has been referred to in Section 4.1. The process has proved to achieve similar yields compared to the conventional catalytic reforming process to produce aromatics but with considerable lower cost of the feedstocks and a huge reduction in the carbon footprint. Hence, one of the main interested parties is the PET industry; therefore, in the long term if this process is successfully implemented in large scale, it may become a risk for the producers of paraxylene, especially if the demand of fuels is reduced over the following decades.

- Hydrolysis and Fermentation

  Alternative routes have been explored aiming for the fermentation of the biomass feedstock to further produce alcohols with a specific number of carbon atoms which can be dehydrated to produce the corresponding olefines (i.e., ethanol, propanol, butanols, butanediols).

4.3.4.3 Dehydrocyclodimerization of LPG

This process is very well known and was designed for its implementation in regions where the availability of naphtha is limited for its conversion into aromatic products. This route achieves almost full conversion of the paraffins; therefore, the separation of reactor effluent is less challenging (i.e., no extractive distillation is required) and contributes to the generation of hydrogen with yields slightly higher than catalytic reforming. However, nowadays the main competitor of this technology is the generation of propylene from propane (i.e., PDH or steam cracking).

4.3.4.4 Shock Wave Pyrolysis

This is one of the most sophisticated methods to produce olefines, and the technology consists of a so-called shock wave reactor which uses a standing shock wave in a supersonic flow of steam to provide the required energy to achieve a temperature pulse tailored to maximize the yield of desired products. This alternative claimed to achieve higher yields of ethylene with a lower energetic consumption and a considerable reduction in coking compared to the conventional steam cracking; however, the steam consumption was up to ten times higher than the typical requirement in a steam cracker; therefore, no further development was attempted.

4.3.4.5 Methanol to Aromatics

This technology is an extension of the methanol to olefin process described in Chapter 2, where the olefins can further be transformed into aromatics using the same zeolite-based catalysts, and similar developments have been explored to produce gasoline from methanol; therefore, the improvements in catalysts to reduce coking and maximize the conversion can at some point be implemented for aromatic production. The commercial implementation of this alternative would open the door for further developments with a positive environmental impact such as the biomass processes described above or even the destruction of carbon monoxide from another sources which can be transformed to methanol through a catalytic process using hydrogen also available from neighbor integrated facilities.

4.3.5 Capital Cost of Integrated Schemes

The capital expense (CapEx) of building or revamping integrated petrochemical-refining facilities will depend on the region, level of integration, and of course capacity of each specific unit. In order to understand the impact of each of these variables, the CapEx can be analyzed based on each of its components according to the following typical breakdown:

- Direct cost materials
    - ISBL
    - Off-sites and utilities
    - Other infrastructure
- Direct cost of labor
    - Indirect cost
    - Feasibility studies
    - Engineering design
    - Procurement
    - Project management
    - Permitting
    - Construction cost
        - Temporary facilities
        - Supervision and management staff
        - Construction equipment
        - Insurance and bonds
    - Commissioning cost
    - Contingency and escalation.

An important point to highlight related to infrastructure is the amount of capex required for the Outside Battery Limits (OSBL); in a typical refinery or petrochemical complex the capex required for the OSBL assets (i.e., storage, handling, utilities, buildings, etc.) can exceed more than 50%–60% of the total capex. This is one of the opportunities for an integrated system to provide value since the integration can have a positive impact (reduction of utilities requirements, shared handling facilities, etc.); however, the overall value will depend also on the location of existing infrastructure in relation to the main processing units (i.e., if the integration is proposed between two sites several miles apart this opportunity may lose some value).

The breakdown shown above includes several items related to indirect cost (i.e., preliminary studies, design, permitting, commissioning, etc.) that may represent a considerable fraction of the capex. This is of importance for an integrated facility, since a considerable amount of effort needs to be invested to ensure that the facilities are designed with enough flexibility to respond to two different market needs sometimes competing with, but sometimes complementing each other.

Another important variable to keep in consideration is the capacity of the proposed units, and most of technologies available are designed to work independently and to perform in competitive scale economy if as a result of an integrated optimization the conclusion is to require units with small capacities (i.e., if enough propylene can be recovered from crackers and FCC units, an on-purpose propylene unit may be needed but at a reduced capacity), such unit may not be able to work efficiently with a consequent impact on the operational cost (i.e., shorter runs between catalyst regeneration for fixed bed processes).

In terms of location, there is no single answer to the impact on capital cost. Normally, these facilities are located in industrial hubs with access to both crude and natural gas pipelines as well as import/export facilities. However, not all these hubs around the world have the same availability of skilled personnel and equipment manufacturing facilities. Therefore, both direct and indirect cost will be completely different for an exact same scheme in the Gulf Coast compared to the West Coast or South Asia, and the impact is so high that interconnection of streams even in long distance may be preferred rather than building an asset in an unfavorable location.

Finally, an important element to consider is the risk, and these schemes may be considered as new developments form the commercial point of view and will require a deep analysis to support a strategy decision at different stages of the project from their construction, commissioning, and operation. Therefore, an increased contingency cost or at least the imposition of several conditions (shorter recovery times, higher insurance costs, etc.) either to execute or to finance the projects may disincentivize these investments to approve their implementation.

4.4 TECHNICAL CHALLENGES AND OPPORTUNITIES

4.4.1 Interfaces during Integration

In order to arrive at a successful integration scheme, different optimization functions must be solved to achieve the different targets for both types of facilities. As described in previous sections, for a refinery the goal is to maximize the production of fuels for a given crude feedstock, while for a petrochemical facility, the target is to produce unsaturated components required as intermediates to produce chemicals for final use. Depending on the location, some feedstocks for the olefin crackers or aromatic trains can be supplied from refining operations such as LPG, naphthas, gasoil, reformate, etc. On the other hand, in the crackers the unsaturation process releases hydrogen and light components highly valuable for some of the refining units.

All these streams will define internal interfaces that depending in their value, and availability may represent specific constraints for an optimization model in the integration process; for example, an integration scheme can be proposed to maximize revenue during more stringent emission requirements in fuels for a given maximum amount of hydrogen available from the crackers to recycle toward refining units. If this interface (the amount of hydrogen available) has a visible impact on the optimization solution, then such interface can be considered as an active constrain and a careful sensibility analysis will give a big picture to propose an integrated scheme.

Therefore, the understanding of the value of these interfacing streams is critical to define optimization functions and the possible ranges of opportunity, and this value will depend on the location and the alternative sources in the area. Coming back to the example of hydrogen, if the available flows of hydrogen from the cracker along with available capacity from steam reformers in the refinery are not enough to satisfy the additional hydrotreating requirements, steam reforming capacity will need to be revamped, and then, this investment needs to be quantified against the value of the hydrogen in the local market as an external supply; in regions with limited availability of natural gas, the model will be impacted by the economics of this alternative source, to identify if there is more value by increasing the recycling of this stream (with a consequent increase on olefins production) or by increasing supply from external sources. This whole scheme will change completely in regions where natural gas is readily available at a more competitive cost.

Not all the streams have the same behavior and therefore will have a different impact on the optimization of an integrated scheme. For example, the value of the naphthas is defined by their trading price referenced to a specific region, i.e., Gulf Coast, Middle East, etc. Therefore, their corresponding constraint in an optimization model may not be as active as any other stream; however, this stream is valuable to produce aromatics and olefins; hence, the distribution to either of these two value chains will have an indirect impact on the integration. Depending on the availability of natural gas to feed the steam crackers, naphtha streams will be preferred as feed for aromatic units which has an impact on the operational mode of the reforming units. But in this case the natural gas value is defining the constraint rather than the value of naphtha since the value of the feedstock is benchmarked against the value of natural gas as feedstock to the crackers.

Another interface is outside of battery limits of these units and is particularly important for integration of multi-location facilities. As discussed in Section 4.3.5, some locations may not be suitable to build a specific unit; however, such facility may be required to either satisfy the local demand of a product or process a by-product already generated from an adjacent unit (i.e., a cracker installed in a region with high availability of natural gas and high demand of ethylene but minimum demand of butadiene generated in a volume where a dedicated unit may not be justified). In this case, the interfaces are defined by offsite infrastructure to interconnect facilities and by being able to move by-products between those. Figure 4.6 shows the infrastructure available in different regions of the USA to interconnect petrochemical and refining facilities.

4.4.2 Planning and Scheduling

Once an optimum integrated scheme has been defined, engineered, and built, another challenge is the short-term planning and scheduling of operations; since now, the system is not only more complex but also different in comparison with the conventional independent operations and at some degree customized for each specific facility.

Both refining and petrochemical operations are very large scale in nature; therefore, a minimum change in the operational mode (yields of desired products, quality of feedstocks, etc.) will have a substantial impact on their margins, and such operational philosophy will be defined according to a predefined commercial strategy which in principle will target for an optimum utilization of the assets; however, in order to materialize this target, a careful and very detailed scheduling strategy needs to be in complete alignment with the corporate strategy and the operational requirements and constraints in the units.

Each company has their own strategies, but in general, planning and scheduling activities are defined with elaborated optimization models as described in Section 4.1 (i.e., linear programming). However, a mathematical model is not always capable of identifying real-time status of all operational variables; for example, in the model, the availability of feedstocks may not represent an issue if continuous upstream supply is guaranteed; however, in reality, the optimization may not have enough information about the availability of supporting infrastructure (i.e., issues on the availability of

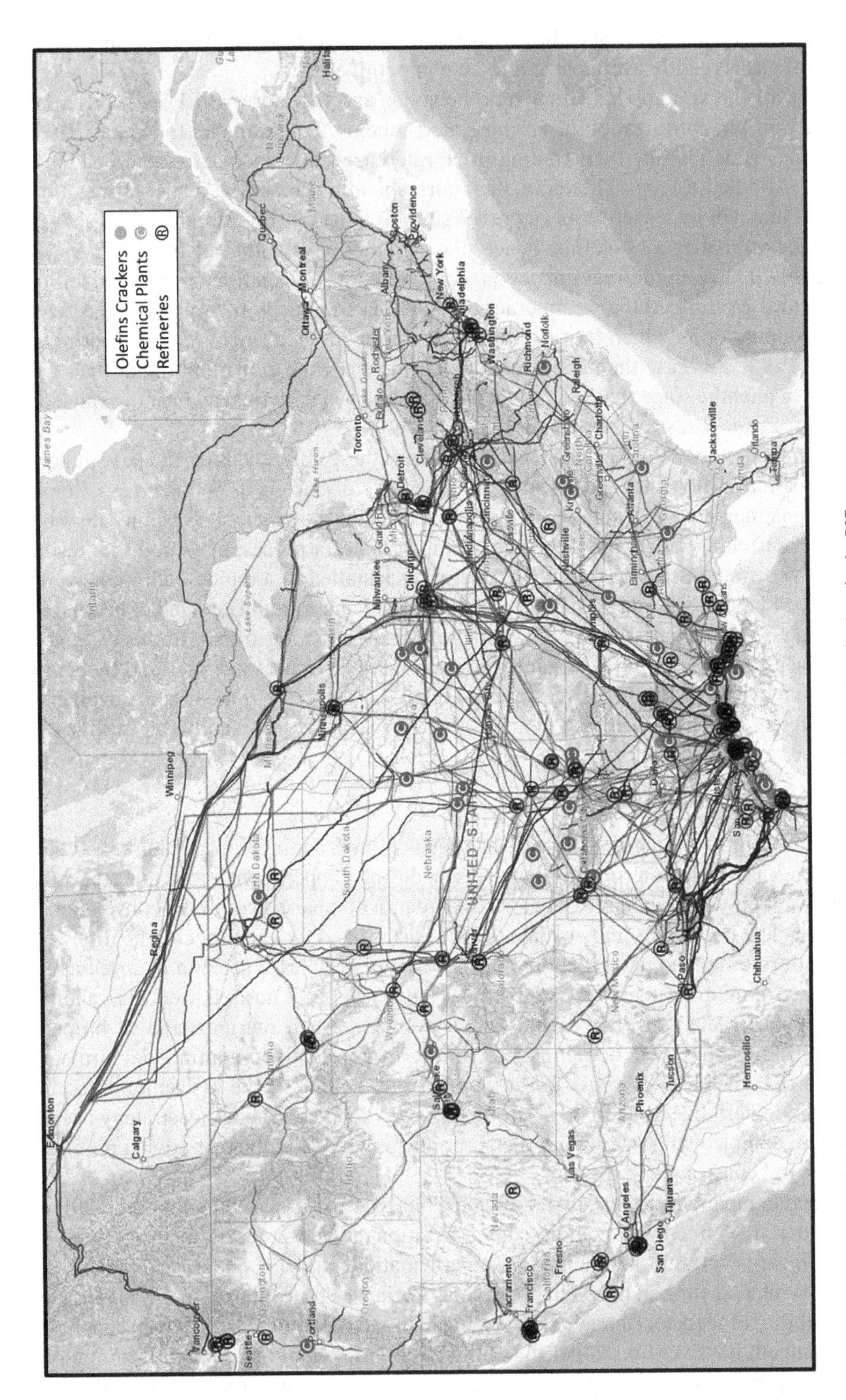

FIGURE 4.6 Interconnecting infrastructure between refineries and petrochemical units in US.

storage to receive the feedstock); therefore, the optimization model will not be able to identify critical bottlenecks, hence this kind of gaps in scheduling programs are normally filled with empirical knowledge, and the priority normally is to use such experience to find the most feasible schedule based on available resources on a day-to-day basis having as a secondary target the optimum operation of the processing units.

An integrated model capable of including process parameters, scheduling data, and commercial targets is very complicated to solve even for isolated petrochemical or refining units; hence to simplify the analysis, these mathematical models normally treat the processing units as black boxes, where heat and material balances are evaluated over wide envelops losing some visibility on internal streams that can provide critical information to find opportunities for improvement (i.e., energy breakdown between unit operations, catalytic regeneration, inefficiencies, etc.). Here is where the operational feedback is critical to improve the accuracy of the models and the identification of parameters with higher sensibility and potential bottlenecks. However, for new integrated and customized schemes, a learning curve needs to be grown before taking advantage of the empirical knowledge, and it will be highly influenced by the rotation of schedulers and operators and the seasonal demands required for these new schemes.

One of the tools recently available is the Deep Learning (DL) which comes from the research on Artificial Intelligence (AI). DL uses algorithms which are capable of solving a given problem as well as can be trained to learn the models without an explicit definition of the problem. The basis of DL is an algorithm called perceptron which mimics the human neuron in the way that it processes information; the key is that such information comes with a specific weight, and then, in order to release this information to another perceptron, the total weighted information has to reach a threshold; otherwise, the process is bypassed, and the algorithm moves to another set of data and repeats the process. As in a normal learning process, the effectiveness of the algorithm will improve not only with the training strategy but also with the quality of the information supplied as input. The power of the perceptrons comes when they are mutually connected (same as in the neurons); therefore, the proposed schemes in DL are based on multiple layers of perceptrons working together to identify complicated patterns and develop the ability of recognizing and predicting data trends which are out of the reach of conventional mathematical and statistical models.

Coming back to the problem of defining planning and scheduling strategies to maximize the value of the integrated assets, one of the challenges is to accelerate the learning curves to acquire empirical knowledge of new process schemes in operation, and here is where the use of new tools such as DL can provide a substantial value to reach a successful implementation of these new process schemes faster, thanks to their ability to classify data based on weight which in reality can be translated on the definition of priorities to achieve optimum performance by predicting possible gaps in the sequence of activities upstream and downstream of the assets.

Another challenge to overcome during the planning process is the volatility on the market for either feedstocks or products. The information available during the design of the process units may change dramatically over the years; a clear example is the volatility on the markets experienced during recent impact in the pandemic period of COVID-19. At early stages of the projects, designers will target for the most flexible range of operation of the processing units (easy interconversion of cracker feedstocks, change of yields on FCC units or reformers, etc.); however, this flexibility has a limit and at some point external factors (environmental restrictions, market conditions, etc.) may push the facilities to run close to those limits; therefore, an ingredient to consider is the probability of reaching these limits during the life of the assets. Then, a continuous risk analysis needs to be conducted to quantify this probability and the corresponding impact day to day to find adequate mitigations or required adjustments either in the facilities or in the optimization functions during the planning process.

4.4.3 Impact on Operational Staff

As discussed in the last section, the role of operational staff has been critical for a successful implementation of integrated facilities. Operational, maintenance, and scheduling staff own all the knowledge required to run the facilities in a safe and efficient way, and they must understand the behavior of the facilities under different situations and must be capable of customizing the process and control systems to ensure an optimal level of performance for a given set of parameters to meet based on a predefined schedule and aligned with restrictions from the maintenance requirements day to day.

This level of performance can only be achieved with proper training and continuous improvement, but the particularity of new integration models is that such training must include a complete vision of the strategy in the company. Operating staff spends several hours a day solving specific and very particular problems in a daily shift, and then over the time, it is very easy for them to lose track of the big picture; they need continuous training and feedback to keep a continuous alignment on the final goals of the company when running these assets; otherwise, they may end treating a fuel gas compressor feeding gas for heaters in the operators' room with the same priority as a hydrogen compressor feeding several hydroprocessing units.

Once the overall corporate strategy has been permeated to all the operational levels, the staff will be more receptive to understanding the level of complexity of the integrated units and therefore will be able to define priorities, and understand not only how to identify opportunities for improvement but also how to anticipate possible areas of risk with impact in the reliability of the assets but most importantly in the safety of the operations.

In order to establish a solid training strategy, a robust information management system is fundamental. Updated operational and maintenance procedures developed through a very well-defined life cycle process will contribute to an even growth of the technical skills of all the operational staff instead of having isolated lots of knowledge segregated over more experienced operators only. This is particularly important when a new integrated asset is operated by operators with experience in similar isolated units because by natural tendency, the staff will try to replicate what used to work in previous experience but now is not exactly applicable (i.e., yields of a reformer running to maximize benzene and other aromatics production may represent a challenge for an operator who for several years has run the units to maximize octane rating on the reformate).

On the other hand as these integration schemes become more popular, one of the challenges will be the lack of skilled personnel to run these facilities; this may increase the need of robust training systems as described above, but once the onboarding process is complete, there is always a latent risk of high rotation, and the companies will need to incorporate efficient strategies to ensure long-term retention of key personnel; otherwise, a high rotation will have a negative impact on the learning curve to achieve the performance level at which the processing schemes were conceived.

Another important role of operating staff when running these integrated facilities is in the definition of efficient sets of metrics not only to compare the performance of these integrations against the actual design parameters vs. previous non-integrated schemes but also to identify potential gaps to prevent malfunctions and reduce downtime of the facilities as well as to find opportunities for improvement and provide feedback to the design of new schemes.

In order to close the cycle, a combined effort from operational staff, strategic teams, management, and projects (i.e., design) is needed to find efficient structures to analyze real-time data collected from operations to improve the overall efficiency of the company not only from the revenue point of view but also keeping in mind environmental, sustainability and safety requirements as a priority.

4.4.4 Safety and Environmental Challenges

Refining and petrochemical units work at operating temperature and pressures very far from normal ambient conditions with very hazardous products; therefore, the equipment and related infrastructure must be carefully designed to avoid any release of materials or energy that can result in the harm of the personnel around the facilities or neighbor communities in the surrounding areas. However, to ensure a safe operation of the facility according to the original design a very strong process safety management system must be in place previous to the operational stage of these facilities.

Such safety management system demands a very clear understanding of the hazards and potential risks during different operational situations of each of the units, which normally comes from a detailed interdisciplinary hazards analysis conducted at different stages of the conception of the facilities (design, construction, commissioning). However, an important ingredient during these sessions is the operational knowledge gathered from previous experience.

A very large compilation of lessons from previous incidents is available for both refining and petrochemical facilities. However, the targeted synergies and interconnections between these industries are somehow a relatively new practice in comparison with the standalone operational mode. Therefore, during the definition of these synergies two challenges need to be overcome in terms of safety: the first is the lack of experience during the integration, and the second is the increase of the complexity in comparison with the original operational philosophies of the units.

Fortunately, given the importance of process safety nowadays, there are several tools and strategies to implement an adequate safety management system which in general is built over a strong safety culture that, in a similar way as with the company strategy, has to be shared at all levels to communicate what is the output expected as a company in terms of safety and how each employee has an important and active role in achieving this goal.

A process safety management system will rely on different building blocks for a successful implementation, but one of the most relevant in integrated refining-petrochemical units is the management of safety knowledge. Building a strong knowledge system will be critical for future facilities over the time, and this opens an opportunity for collaboration to standardize how this knowledge is transferred and what are the most critical areas of focus which may have a positive impact on improving the output of hazard analysis, management of change process as well as the structure of operational procedures.

With the same level of importance as safety, the impact to the environment is another priority during the design construction and operation of any processing facility, and the good news of integrated units is that the environmental impact is one of the drivers for the synergies. The integration promotes the sharing of supporting infrastructure such as utilities and storage; therefore, an optimized use of such infrastructure will immediately be translated into a positive environmental impact with substantial reduction on direct emissions.

However, there are two main threats that will drive the future of these facilities and have a close connection with their indirect environmental impact. One is the evolution of renewable fuels, and it may be too early to have a conclusion, but it is possible to see this integrated schemes between refineries and petrochemical units as a gradual fading of the fossil fuel industry, and such integration is no other than a hand over process of the assets that in some decades may no longer be needed to produce fuels.

The second issue is the public perception of the petrochemical industry as the main producer of single-use plastics; in the last years, a big debate has been in place due to the negative impact that is claimed these plastics have on the environment and ecosystems; however, the petrochemical industry is also the source of feedstocks to produce a wide range of final products that are fundamental for our day-to-day activities, computers, medical equipment, detergents, lubricants, drugs, etc. Therefore, eventually we need to find creative solutions to mitigate the negative impact of this industry because at the end, it is one of the cornerstones of the development achieved as a modern society.

BIBLIOGRAPHY

Amghizar I.; Vandewalle L.A.; Van Geem K.M.; Marin G.B.; New trends in olefin production, *Engineering*; March 2017.

Baroi C.; Gaffney A.M.; Fushimi R.; Process economics and safety considerations for the oxidative dehydrogenation of ethane using the M1 catalyst; May 2017.

Energy Efficiency Improvement and Cost Saving Opportunities for the Petrochemical Industry; An ENERGY STAR® Guide for Energy and Plant Managers; Maarten Neelis, Ernst Worrell, and Eric Masanet; Enrest Orlando Lawrence Berkeley National Laboratory; June 2008.

Gas Technology Institute; Refinery upgrading of hydropyrolysis oil from biomass; 2015.
Gupta A.; Introduction to deep learning, *Chemical Engineering Progress*; June 2018.
Held A.; Production of Renewable Aromatic Chemicals using Virent's Catalytic BioForming® Process; October 2010.
Hydrogen Management in Refineries; *Petroleum & Coal; Zahra Rabiei; Oil & Chemical Engineering Development*, Sharif University of Technology, Tehran, Iran; November 2012.
Parthasarathi R.S; Alabduljabbar S.S.; HS-FCC High-severity fluidized catalytic cracking: A newcomer to the FCC family; September 2014.
Petrochemicals Europe; Petrochemicals make things happen; PE flowchart updated 2015; Version 031.
Refining – Petrochemical Integration; Fluor; Claus-Peter Haelsig, Fred Baars; Egypt Downstream Summit & Exhibition; 2016.
Rethinking the Water Cycle; McKinsey & Company; Martin Stuchtey; May 2015.
Synergies between Refining and Petrochemicals: Today and Tomorrow; Total Raffinaderij Antwerpen; February 2008.
Thailand Industry Outlook 2017–2019; Petrochemical Industry, Krungsri Research; April 2017.
Using Advanced Analytics to Boost Productivity and Profitability in Chemical Manufacturing; McKinsey & Company; Valerio Dilda, Lapo Mori, Olivier Noterdaeme, and Joris Van Niel; January 2018.

5 Conclusions

As per the true definition of the word synergy, leveraging the synergies of the integration between petroleum refining and petrochemicals will create an organization that will have a greater value than the simple sum of the value of the two organizations.

Being able to leverage the synergies of the integration requires self-respect within the both entities, respect for each other and for the competition, open, honest, and effective communication, and a passion for the ultimate goal: being successful in the business and growing the bottom line.

Both industries – petroleum refining and petrochemicals – have a long history of failures and victories. Both show impressive developments and innovation in technologies as they were the enabling industries supporting the lifestyle, we all became accustomed to. Our lives would be a lot less comfortable without the fuels that allow private transportation and transportation of goods, without the chemicals that keep us healthy, and without the materials that make consumer goods and electronics affordable for most of the planet's population. Significant improvements were made in the areas of safety, environmental compliance, product output, and product quality. However, these improvements will not be enough for companies in each of the industries to achieve their respective growth targets, or even to survive in the future market and regulatory conditions.

The refining industry faces a multitude of challenges in the next 15 years that can be summarized as follows:

- Declining demand for fuels in fully developed economies such as Western Europe
- Increasing cost pressure due to further reduction in sulfur content of fuels
- Global decarbonization of fuels to meet targets for reducing carbon dioxide emissions
- Rising market share for renewable fuels and alternative fuels from low carbon sources
- Regional over-capacity and declining refinery utilization.

The dominant theme for the refining industry is defining its role in the decarbonization of the energy and fuel sector. There is no doubt that refineries will still be needed in 2035 and far beyond that point to support the manufacturing of the large number of products that require molecules derived from petroleum. But there is also no doubt that the energy sector must make the shift from fossil fuels with high carbon intensity (mainly coal and petroleum) via a lower carbon intensity source such as natural gas to a zero-carbon or carbon-neutral future. This will be the main driver for a change in orientation and the need to increase the level of integration between refining and petrochemicals. Other measures to bridge the transition time to develop and implement carbon-neutral technologies and help in achieving emission targets include

- Electrification of short-range transportation
- Electrification of heating demand (residential and industrial)
- Carbon capture, utilization and storage (CCUS)
- Continued improvements in fuel quality and usage efficiency.

Concerning a forecast for the energy and fuel demand, COVID-19 has taught us that any prediction of demand developments is just a best guess and will change constantly based on events that are unpredictable and out of our control. Near-term growth drivers will be the developing economies in Asia, mainly China and India. Of course, a country such as China will take advantage of the cheapest form of energy available to manage demand growth. In this case, most of the energy and fuel demand as well as supply of products for the chemical industry is met using coal. South America is struggling with internal turmoil, especially in Venezuela, but to some degree also in countries such as Brazil and Argentina. So, this region is a bit of a wild card when it comes to economic growth and demand development. And the African countries can be considered the sleeping beauty of the economies. There are initiatives that indicate that those African countries with known energy resources such as Nigeria, Cameroon, and Mozambique may become the next developing economies driving growth in the energy sector and other industries.

And the Middle East will continue to solidify its role as supplier of primary energy sources as well as intermediate products to growing economies.

The petrochemical industry faces different challenges over the next 15 years:

- Increasing competition as additional capacities come online along the US Gulf Coast (ethane-based steam cracking), in the Middle East and in Asia (coal- and naphtha-based manufacturing)
- Compensating the historically cyclic demand curves to stabilize the markets, for example, by having greater feedstock flexibility
- Adjusting to changes driven by environmental concerns, for example, implementation of a closed-loop, circular economy for plastics to reduce plastic waste in landfills and oceans
- Finding outlets to monetize by-products from the manufacturing processes.

The dominant theme for the petrochemical industry in the next 15 years will be feedstock supply and staying competitive on markets that will attract more and more players in growth regions with abundance of cheap energy and feedstock supply. Naphtha is the main feed stream feeding the petrochemical plants in Western Europe and Asia. The plants in the Middle East consume a cocktail of almost even parts of ethane, naphtha, and other oils. And the North American/US Gulf Coast capacities are built on ethane supply. Feedstock type and quality determines the yields coming from the steam cracker and the available value chains that can be pursued with the products from the steam cracker. The economic growth in Asian countries such as China has allowed petrochemical plants in the rest of the world to expand their market reach and export petrochemicals and chemicals in support of the growing demand in Asia. As the domestic industry develops and catches up, these export avenues for other countries will get difficult to maintain as they be challenged to compete with the local manufacturers. The fight for operational excellence and cost efficiency will be decided by the consumer who will determine how much money they are willing to spend for the consumer goods and electronics and other products coming out of the petrochemical industry.

The obvious path forward to address the challenges both in the refining and in the petrochemical industries is the path of integration, which brings the following advantages:

- Alternative use for fossil fuels as feed streams to petrochemical plants
- Feedstock flexibility for the petrochemical side
- Use of excess hydrogen from petrochemicals in refinery processes
- Use of common infrastructure
- Use of common utilities
- Lower OPEX due to increased operational efficiency
- Lower fixed cost through use of common administrative functions
- Use of petrochemical by-products as feed or blend stock in the refinery
- Increased flexibility to react to market shifts and increased competitiveness.

The challenges for integration of refining and petrochemicals lay within the increased level of complexity of the combined facilities and the difficulty of managing and optimizing such a complex system. Luckily, advancements in available technologies such as affordable computing power and improved software tools enable companies to handle this challenge by use of big data analysis, the Industrial Internet of Things, improved process simulation, and enhanced advanced control systems. Implementation of these tools will enable organizations to properly plan feedstock supply and production rates, monitor and control operations, and evaluate ideas for optimization of any of the processes.

In parallel, processing technologies will develop to improve the outcome of an integration between refining and petrochemicals, for example:

- Oxidative dehydrogenation
- Biomass gasification or catalytic pyrolysis of biomass
- Dehydrocyclodimerization of LPG
- Shockwave pyrolysis
- Methanol to aromatics.

Product exchange between the refinery and the petrochemical plants will include naphtha, C_4 streams, olefins, ethane, methane, aromatics (BTX), and hydrogen. The integrated complex will also be capable of utilizing renewable energy sources to reduce their carbon footprint, take advantage of increased energy efficiencies, and improve their image in the public. The Middle East will establish itself as the processing hub for fuels and petrochemicals to support the growing economies in Asia. And products from Asia will be exported to the established economies in Western Europe and North America.

Index

Taylor & Francis eBooks

www.taylorfrancis.com

A single destination for eBooks from Taylor & Francis with increased functionality and an improved user experience to meet the needs of our customers.

90,000+ eBooks of award-winning academic content in Humanities, Social Science, Science, Technology, Engineering, and Medical written by a global network of editors and authors.

TAYLOR & FRANCIS EBOOKS OFFERS:

A streamlined experience for our library customers

A single point of discovery for all of our eBook content

Improved search and discovery of content at both book and chapter level

REQUEST A FREE TRIAL

support@taylorfrancis.com

Printed in the United States
By Bookmasters